AF670444

Imaging in Nuclear Medicine

Augusto Giussani • Christoph Hoeschen
Editors

Imaging in Nuclear Medicine

Springer

Editors
Augusto Giussani
Department of Radiation Protection and Health
BfS – Federal Office for Radiation Protection
Oberschleißheim
Germany

Christoph Hoeschen
Research Unit Medical Radiation Physics and Diagnostics
Helmholtz Zentrum München – German Research Center for Environmental Health
Neuherberg
Germany

ISBN 978-3-642-31414-8 ISBN 978-3-642-31415-5 (eBook)
DOI 10.1007/978-3-642-31415-5
Springer Heidelberg New York Dordrecht London

Library of Congress Control Number: 2013933985

Printed on acid-free paper

Springer is part of Springer Science+Business Media (www.springer.com)

Contents

Chapter 1
Introduction

Augusto Giussani and Christoph Hoeschen

Nuclear medicine is gaining greater and greater importance as a diagnostic tool, and one reason for that is the enormous progress made in the imaging of the tissue activity distribution and of its temporal variation after administration of radiolabeled agents to patients. The availability of hybrid systems, where SPECT or PET machines are combined with X-ray computed tomography (CT) or, more recently with magnetic resonance (MR) devices, enables to integrate together the functional information provided by the administered radiopharmaceutical with the superior anatomical details of CT and MR. A look at the figures that are reproduced in the chapters of this book (see in particular Chap. 2) gives a clear indication on how the imaging processes have evolved over the years.

The generation of the image in nuclear medicine starts with the detection of the radiation emitted by the administered radioactive agent that has accumulated in specific tissues, including malignancies, and includes signal analysis and processing, noise reduction, image reconstruction, image representation, maybe even image analysis, and so on. All these aspects have profited over the years of the results of research activities conducted in the field of nuclear medical imaging and also in related fields, and even when those activities were not directly aimed to nuclear medicine, their results could be more or less easily implemented. This is valid for the development of new detectors, new detection setups, new

A. Giussani (✉)
Research Unit Medical Radiation Physics and Diagnostics, Helmholtz Zentrum München – German Research Center for Environmental Health, Ingolstädter Landstr. 1, 85764, Neuherberg, Germany

Department of Radiation Protection and Health, BfS – Federal Office for Radiation Protection, Ingolstädter Landstr. 1, 85764, Oberschleißheim, Germany
e-mail: agiussani@bfs.de

C. Hoeschen
Research Unit Medical Radiation Physics and Diagnostics, Helmholtz Zentrum München – German Research Center for Environmental Health, Ingolstädter Landstr. 1, 85764, Neuherberg, Germany

A. Giussani and C. Hoeschen (eds.), *Imaging in Nuclear Medicine*,
DOI 10.1007/978-3-642-31415-5_1, © Springer-Verlag Berlin Heidelberg 2013

mathematical approaches and algorithms for signal processing and reconstruction techniques, not to mention the booming of the computer technologies and the increased performances of the processors used for reconstruction and representation or analysis tasks like 3D image representation.

In this book, which finds its origin in a Training Course held in Munich in October 2010 (Third MADEIRA Training Course on Imaging in Nuclear Medicine), we try to give an overview of the recent and future trends in the fields that did and will contribute to the process of imaging using radioactive imaging agents.

Part I gives a general overview of imaging in nuclear medicine. The book starts with a medical perspective: in Chap. 2, Professor Astrid Riklund, chief physician at the Department of Radiation Sciences of the University of Umeå, Sweden, describes the current and future challenges to nuclear medicine imaging from the point of view of the "end users," i.e., the medical doctors that have to evaluate the images. In Chaps. 3 and 4, Andrej Studen, from the Jožef Stefan Institute in Ljubljana, Slovenia, and Axel Martinez Möller, from the Technische Universität München, Munich, Germany, describe, mainly from a physical point of view, the processes related to the collection and formation of the electrical signals which are the inputs for the generation of the images. The principle of tomographic and hybrid systems, by now representing the large majority of devices in use, is described there. Chapter 5 goes one step beyond it. Stephan Walrand and François Jamar, from the Université Catholique de Louvain, Brussels, Belgium, illustrate the newest developments and futuristic concepts for applications in nuclear medicine imaging.

Part II of this book is focussed on the issue of tomographic image reconstruction and processing. The development of new, more efficient algorithms not only leads to an improved image quality and to a shorter processing time but also to a reduction of the activity that needs to be administered to the patient and consequently to a reduction of her/his radiation exposure. The process of image reconstruction in nuclear medicine (which is a so-called emission tomography technique, since the reconstruction process starts with the collection of the radioactivity emitted by the patient) is in several aspects different from the case of "transmission tomography" techniques like CT, where the radiation source is external to the patient and the image is reconstructed on the basis of the transmitted radiation which can traverse the patient's body. However, there are certainly also common approaches as, e.g., in noise reduction principles. Oleg Tischenko, Christoph Hoeschen (both from Helmholtz Zentrum München, Munich, Germany), and Gernot Ebel (Scivis wissenschaftliche Bildverarbeitung GmbH, Göttingen, Germany) describe in Chaps. 6–8 the newest approaches, partially taken from the field of transmission radiography and specifically adapted to the characteristics of nuclear medicine. This section integrates what was already presented in a previous book of this series[1] and shows the newest further developments. Also the new trends in quality assurance in nuclear medicine are described by Hugo de las Heras Gala in Chap. 9. As nowadays the majority of all new equipment consists in hybrid

[1] Cantone MC, Hoeschen C (eds) (2011) Radiation physics for nuclear medicine. Springer, Heidelberg.

systems, quality assurance programs for transmission, as well as for emission imaging are dealt with in this chapter.

Part III of this book is closely related to the outcomes of the European research project MADEIRA (Minimizing Activities and Doses by Enhancing Imaging quality in Radiopharmaceutical Administration). It was in the course of this project, co-funded by the European Commission through the EURATOM Seventh Framework Program, that the Training Course was organized and the book was conceived. Aim of the project, as its title reveals, was to develop several strategies to improve the imaging process and thus enable to reduce radiation dose to the patients keeping the diagnostic information of the image at the same level, or even enhancing it. Chapters 10 and 11 describe, among others, the development of a new detector concept, aimed to provide higher definition images of selected regions of interest and of a dedicated phantom, specifically developed for the experimental determination of the quantities closely related to the image quality.

Finally, Chap. 12 summarizes useful "rules of the thumb" that can be directly put in use for producing and keeping good quality images.

One final remark has to be made: nuclear medicine is likely the scientific and medical discipline requiring the collaboration of experts from the highest number of different fields: physicians, pharmacologists, biologists, chemists, physicists, mathematicians, and programmers. All these experts are concerned by the development of the perspectives presented here. Thus, the texts were formulated with the intention to be understandable by the widest possible audience. As a result, some explanations may appear superfluous or simplistic for some readers. In that case, the reference lists at the end of each chapter are useful tools for deepening the knowledge.

Part I
Basic Principles and Physics of Nuclear Medical Imaging

Chapter 2
The Role of Imaging in Nuclear Medicine: The Medical Perspective

Katrine Riklund

2.1 Introduction

The era of clinical nuclear medicine started already 70 years ago with the first human examination of the thyroid (Fig. 2.1). Iodine's natural affinity for the gland was used for these first examinations in patients. In 1938 the first paper on the diagnostic uses of ^{131}I in thyroid diseases in humans was published [1]. The development of this discipline has thereafter been tremendous, and still we have not seen the end of progress. Nuclear medicine has a large input in the clinical handling of patients with different diseases and is a modality to examine different physiological function in humans. It is a molecular imaging method based on emission gain by injecting the patients with a very small amount of a radioactive radiopharmaceutical or tracer. The distribution is measured with a gamma camera, a single-photon emission computed tomography (SPECT) camera or a positron emission tomography (PET) camera. In gamma and SPECT camera, radionuclides with gamma decay are used and in PET camera with positron emission.

The clinical scene in nuclear medicine today is hybrid imaging. It is the modality developing most rapidly. By hybrid imaging a combination of two different modalities is meant, most commonly SPECT and computed tomography (CT) or PET and CT, resulting in a combination of molecular/biologic and structural information displayed in a fused mode (Fig. 2.2).

Hybrid systems with a combination of PET and magnetic resonance tomography (MR) are being introduced in the clinical setting, and greater details will be given in Chap. 4. The structural part, the CT, contributes with morphologic information but also with transmission data which is used for attenuation correction of the PET or SPECT data. The strengths of CT are the spatial resolution and the contrast [2, 3]. PET and SPECT on the contrary have a relatively low spatial resolution, and the

K. Riklund (✉)
Diagnostic Radiology, Department of Radiation Sciences, Umeå University, Umeå, Sweden
e-mail: Katrine.riklund.ahlstrom@diagrad.umu.se

A. Giussani and C. Hoeschen (eds.), *Imaging in Nuclear Medicine*,
DOI 10.1007/978-3-642-31415-5_2, © Springer-Verlag Berlin Heidelberg 2013

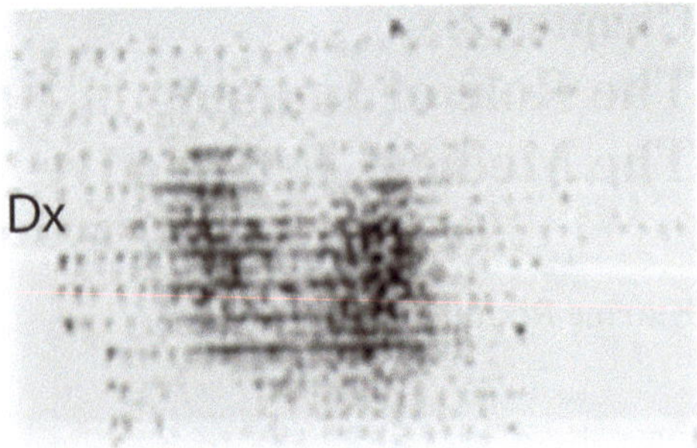

Fig. 2.1 An old image showing the distribution of function in a normal thyroid gland measured with ^{131}I

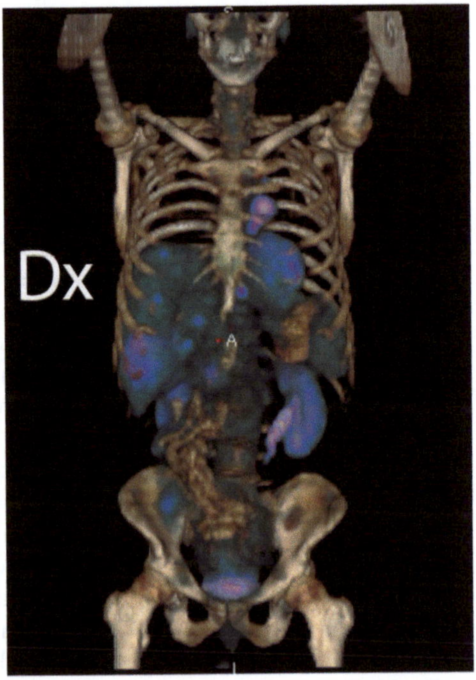

Fig. 2.2 An FDG-PET/CT 3D image of a patient with lung cancer. The skeleton is from the CT acquisition and the uptake in the liver, kidneys, bladder, heart and the lung tumour is from the PET acquisition

strength is the very high molar sensitivity giving the possibility to examine small changes in physiology [4, 5]. A comparison of different imaging modalities is shown in Table 2.1. Also gadolinium-enhanced magnetic resonance imaging (Gd-MRI), which is not based on the use of ionizing radiation, is shown for comparison.

Before an examination with radiation is done, it has to be justified. The appropriateness and the benefit versus risk have to be decided, and the examination procedure has always to be optimized. A radiopharmacon, usually consisting of a radionuclide bound to an active substance defining which function is examined, is needed. For SPECT studies the most common radionuclide is ^{99m}Tc, a pure gamma emitter and 6 h half-life ($T_{1/2}$). For PET the working horse is ^{18}F, a positron emitter

Table 2.1 Molar sensitivity and spatial resolution for different tomographic modalities

Modality	Concentration at target	Resolution (mm)
PET	Picomolar (10^{-12})	4–6
SPECT	Nanomolar (10^{-9})	4–12
Gd-MRI	Milli-micromolar ($10^{-3/-6}$)	<0.5
CE CT	Millimolar (10^{-3})	<0.5

PET positron emission tomography, *SPECT* single-photon emission computed tomography, *Gd-MRI* gadolinium-enhanced magnetic resonance imaging, *CE CT* contrast-enhanced computed tomography

with 109 min $T_{1/2}$. The active substance can be a bone seeker, such as methyl-diphosphonate (MDP) for SPECT or fluro-2-deoxy-2-D-glucose (FDG), which measures the metabolism in PET. The radiopharmacon or tracer can be administered to the patient in different ways, of which intravenous (iv) injection is the most common.

Nuclear medicine covers a variety of clinical questions, and it is the clinical question deciding the setup of a nuclear medicine examination. Some examples of the most common and most recent applications of nuclear medicine imaging, listed according to the imaging modality, are presented in the following sections, and some indications on future trends and need, viewed from the medical perspective, are given in the end.

2.2 Gamma Camera and SPECT Imaging

Examination of the whole skeleton with ^{99m}Tc-MDP (methyl-diphosphonate) to diagnose metastases or other bone diseases is a very common examination with conventional gamma camera (Fig. 2.3). Bone scintigraphy has been the method of choice to follow patients with oncology diseases, but imaging with different tracers in PET/CT is rapidly developing [6–8].

By using a substance measuring the regional brain blood flow (rCBF), the brain function can be measured by a SPECT or a SPECT/CT camera [9, 10]. Hexamethyl-propyleneamine oxime (HMPAO) or Neurolite is distributed in the brain in relation to the blood flow, and most of the uptake takes place during the first passage. The rational for examination is that brain function is correlated to blood flow. The examination is used to diagnose dementia and to differ between neurodegenerative and vascular dementias (Fig. 2.4a). The same substance (HMPAO) can also be used to label white blood cells and with that substance inflammatory or infectious diagnoses can be made. This is mostly used to diagnose inflammatory bowel disorders (Fig. 2.4b). Blood flow and myocardial function can be examined with other radiopharmaceuticals. A cocaine analogue, ^{123}I-FP-CIT binding to dopaminergic neurons, is used to examine the function in striatum which is affected in parkinsonian diseases and Lewy body dementia, but not in essential tremor, making it possible to set correct diagnosis with high sensitivity and specificity (Fig. 2.5) [11–13]. If the patient is allowed to breathe an aerosol of carbon particles covered with ^{99m}Tc, the distribution of air in the lungs can be made.

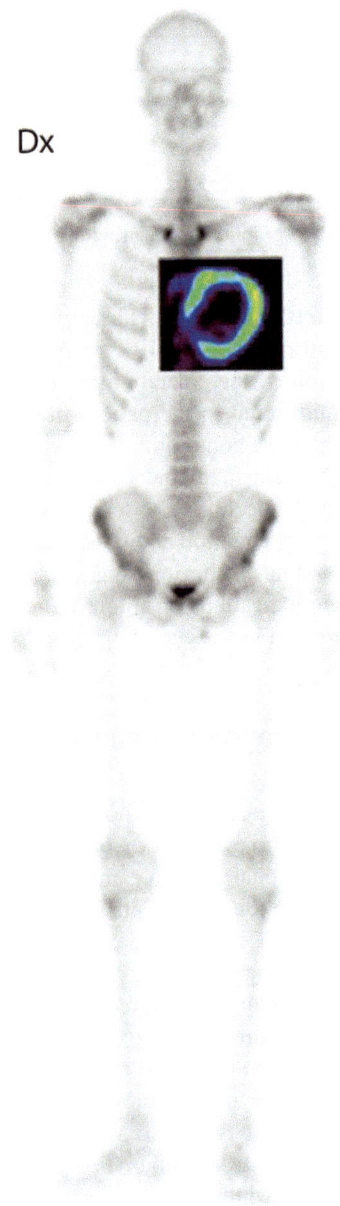

Fig. 2.3 A whole body bone scintigram of a normal skeleton

This is used in combination with an iv injection of ^{99m}Tc-labelled macro-aggregates of albumin, small enough to enter the capillaries. By measuring the ventilation and blood flow distribution in the lungs, pulmonary embolism can be diagnosed. A mismatch is seen in pulmonary embolism, with activity defect in the

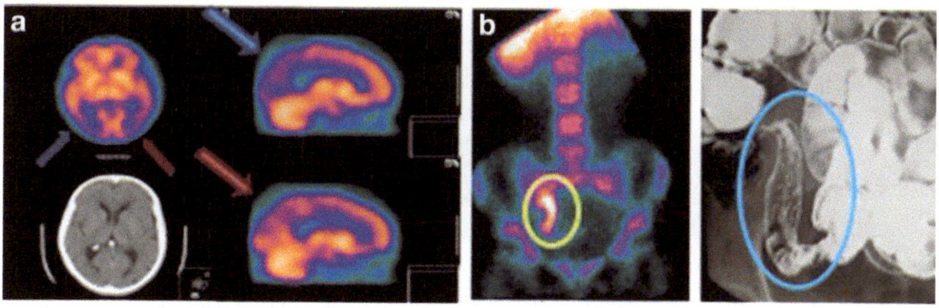

Fig. 2.4 (**a**) A CT and an rCBF-SPECT study with HMPAO of a patient with dementia. The blood flow is reduced in the parietal lobes bilateral (*red* and *blue arrows*). The CT shows no atrophy. (**b**) On the *left*, an intense uptake of HMPAO-labelled white blood cells is seen in the right fossa (in the *yellow circle*) in a patient under investigation of inflammatory bowel disease. The uptake is located to the bowel section with acute inflammation. On the *right*, the cobble stone appearance of the mucosa is visualized with the small bowel contrast examination

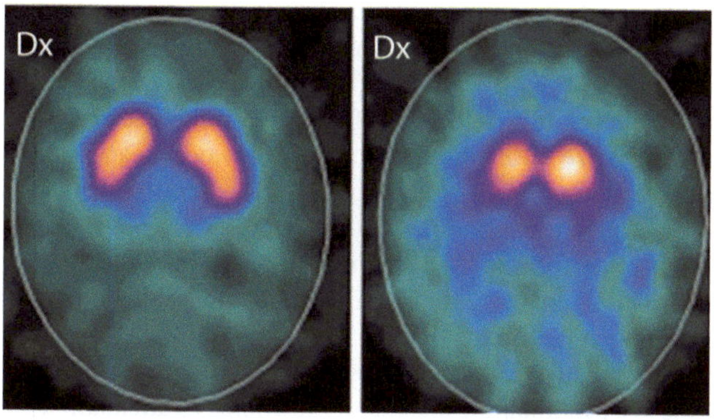

Fig. 2.5 FP-CIT-SPECT images of two patients with tremor. The one on the *left* has a normal dopamine transporter SPECT study with preserved activity in striatum bilateral and is suffering from essential tremor. The patient on the right has reduced activity in putamen bilateral (the lower part of striatum) and on the right nucleus caudatus consistent with a parkinsonian disease

perfusion but not in the ventilation study (Fig. 2.6). Even if CT examination of suspected pulmonary emboli is the first choice of examination, there is a clinical need of scintigraphy studies [14, 15].

2.3 PET/CT Imaging

PET/CT was introduced to the clinical practice in the beginning of 2000 [16, 17] and after that a tremendous development has been taking place. Compared to SPECT, the PET scanner has a significantly higher sensitivity and a slightly better spatial resolution (Table 2.1). Instead of using gamma emitters, positron emitters are used. A positron emitter gives away a positron from the nucleus, and since this antiparticle cannot exist, it interacts with an electron and annihilation takes place.

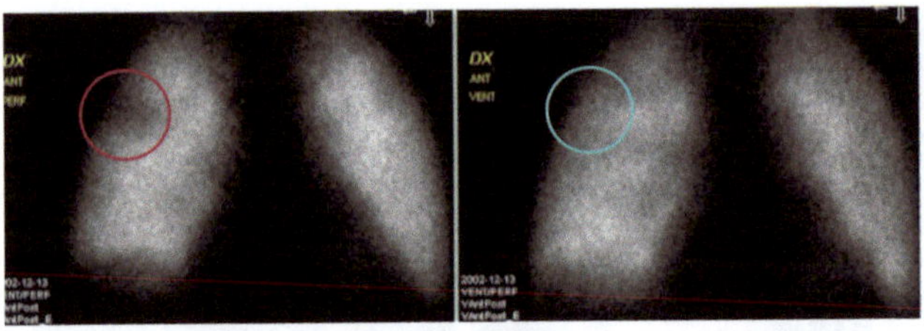

Fig. 2.6 Shows a perfusion and ventilation scintigraphy of a lung. To the *left* an image of a perfusion scintigraphy is seen. A triangular shaped region with reduced perfusion is seen in the upper lateral segment of the lung (*red circle*). To the *right* the ventilation scintigram of the same patient is shown. The ventilation in the entire lung is normal. The corresponding segment to that with reduced perfusion is delineated with a *blue circle*.

The result is that mass converts to energy, and 511 keV photons radiating almost 180° in opposite direction are produced. These 511 keV photons are measured by the PET-detectors and transformed into images, as described in Chaps. 3 and 4. The invention of PET scanning is not new, and already in the 1950s, arsenic was used to image brain tumours [18].

The combination of PET and CT (PET/CT) is a molecular imaging modality allowing us to visualize different biochemical pathways in cancer cells. Imaging is usually done with ^{11}C- or ^{18}F-labelled tracers. The sensitivity is high, making it possible to detect picomolar and down to nanomolar concentrations of the tracer in vivo and thus to perform imaging using non-pharmacological concentrations of the substances. This counts also for SPECT and gamma camera imaging. Oncology diseases are still the major indication for PET/CT even if the use for cardiac, brain imaging and inflammation is increasing.

Initially, the CT examination was mainly restricted to attenuation correction and anatomical mapping to facilitate the localization of the PET uptake. Today, the vast majority of PET/CT scanners are equipped with a state-of-the-art CT, and an increasing proportion of PET/CT studies are made with high-quality diagnostic CT. This true hybrid imaging sets demands on the physicians to have qualification in both radiology and nuclear medicine to evaluate the entire hybrid examination. Collaboration between a radiologist and a nuclear medicine specialist making the evaluation together is another working strategy. Furthermore, the competence of the nurses/technicians is changing, and knowledge from both radiology and nuclear medicine is needed. The clinical routine to examine patients will also change with replacement from CT to PET/CT in an increasing proportion of, for example, lung cancer patients. In summary, the development and diffusion of hybrid imaging has increased the collaboration between radiology and nuclear medicine.

2.3.1 *Fluoro-2-Deoxy-2-D-Glucose*

The absolute most commonly used PET tracer for imaging is FDG labelled with ^{18}F for oncologic diseases. The tracer is delivered by the blood and taken up by the tumour cells with help of glucose transporters. Inside the cell FDG is

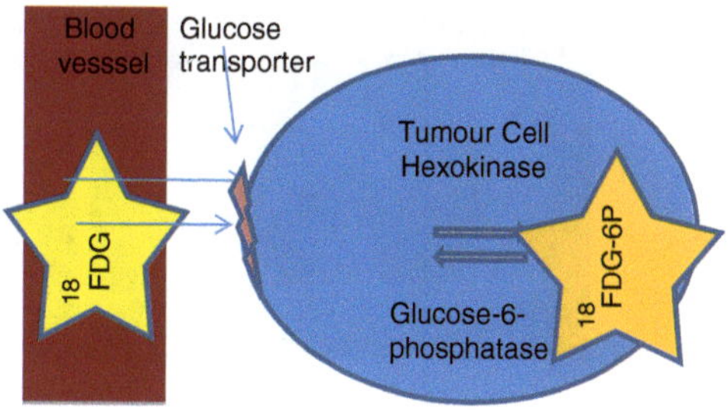

Fig. 2.7 A schematic drawing of the FDG uptake and entrapment in tumour cells

phosphorylated by hexokinase, and the product FDG-6P does not continue in the metabolic route. It remains entrapped in the tumour cell where it can be measured by the PET scanner. Therefore, this tracer acts as a surrogate marker for metabolism, and this pathway is the one which is most frequently imaged. FDG has been in use since 1978 [19], and its use is reimbursed for most solid tumours (Fig. 2.7).

PET/CT with FDG has a high impact in clinical handling, and in 2001, a study was published showing changed clinical management of approximately a third of patients with lung cancer, colorectal cancer, melanoma and lymphoma after adding FDG-PET for staging [20]. In two publications from 2008, the impact of PET/CT on expected management of patients with cancer was evaluated based on ~23,000 PET/CT studies. These papers also concluded major changes in patient handling in approximately one-third of patients with varying oncologic diagnosis. These changes affected clinical handling for diagnosis, initial staging, restaging as well as for suspected recurrence [21, 22]. As medicines become more and more personalized, PET/CT can be of help in staging, in treatment planning and in some diagnoses also in early treatment evaluation.

2.3.2 *Lung Cancer*

There is a strong evidence for the use of FDG-PET/CT in initial staging for lung cancer when curative treatment is intended [23–25]. Evaluation of lymph nodes in mediastinum has the highest added value. If lymph nodes are PET negative, spread disease can be ruled out. FDG-positive nodes on the contrary are recommended to be biopsied before final staging, due to the risk of false-positive findings in inflammation. The staging with PET/CT gives major information for the treatment decision and gives also an indication of prognosis for the patient. For lung cancer, FDG-PET/CT is furthermore recommended for radiation treatment planning in order to include all active disease and to avoid radiation of for instance lung atelectasis which can be impossible to delineate with CT [24, 26].

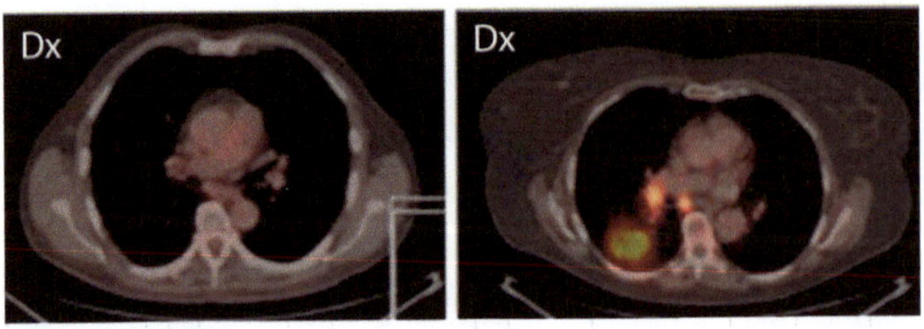

Fig. 2.8 To the *left*, an FDG-PET/CT image of a patient with lung cancer and FDG-negative border-sized lymph nodes inferior of carina. The patient underwent curatively intended therapy and is free of recurrence. To the *right*, an FDG-PET/CT image of a patient with a similar sized lung cancer and border-sized FDG-positive mediastinal lymph nodes. Curative cytoradiation was given. The patient survived 1 year after treatment and died of the disease

2.3.3 Lymphoma

FDG-PET/CT is the best modality to evaluate effect of treatment [27]. Previously only CT was used for response assessment, but recently revised Integrated International Workshop Criteria (IWC + PET) were proposed by the members of the International Harmonization Project [28]. The revised criteria combine both imaging techniques. If lymph nodes which were initially FDG avid do not show FDG uptake after treatment, the patient is considered to be in complete response, irrespective of the size of lymph nodes at CT [29]. The strength with FDG-PET/CT in lymphoma is the possibility to distinguish viable lymphomas from remnants such as necrosis or fibrosis, which are not malignant. FDG-PET/CT is also used for evaluation early in the treatment course, after two cycles of cytotoxic drugs, but more evidence is needed before the clinical role is decided [30] (Fig. 2.8). A rapid decline in FDG uptake indicates good prognosis irrespective of the CT findings. For clinical use visual assessment is used, and criteria for evaluation are published by Cheson et al. and Juweid et al. in 2008 [31, 32] (Fig. 2.9).

2.3.4 Inflammation

Beside clinical tests, structural imaging is used to diagnose vasculitis in the large arteries. It might be hard to decide correct diagnosis since half of the patients with structural findings does not have active inflammatory disease [33]. A FDG-PET/CT can help to solve the diagnostic problem in patients not on steroid treatment, and if positive, a high correlation with disease is found [34, 35] as shown in Fig. 2.10. FDG-PET/CT can also be used to diagnose infection in vessel grafts or abscesses [36, 37].

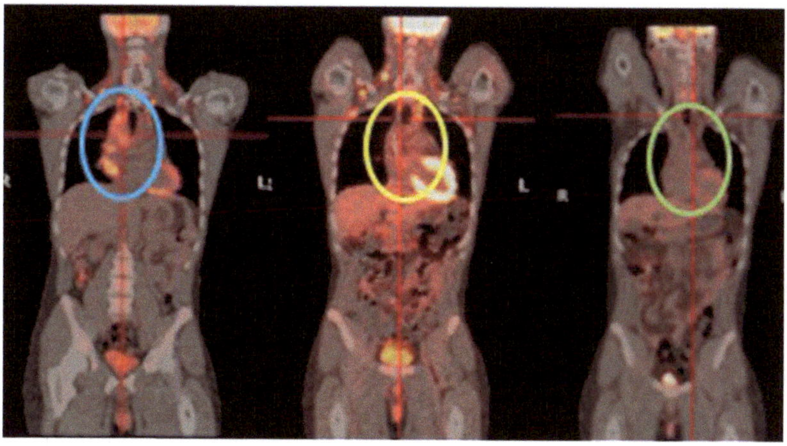

Fig. 2.9 FDG-PET/CT images of a patient with Hodgkin's disease. The image to the *left* is before treatment, the one in the middle after two treatment cycles and to the *right* after treatment is completed. Initially massive uptake in the lymphoma is seen (*blue circle*). After two circles of cytotoxic drugs, most of the lymphoma is FDG negative (*yellow circle*). After completed treatment, all lymphoma remnants are FDG negative

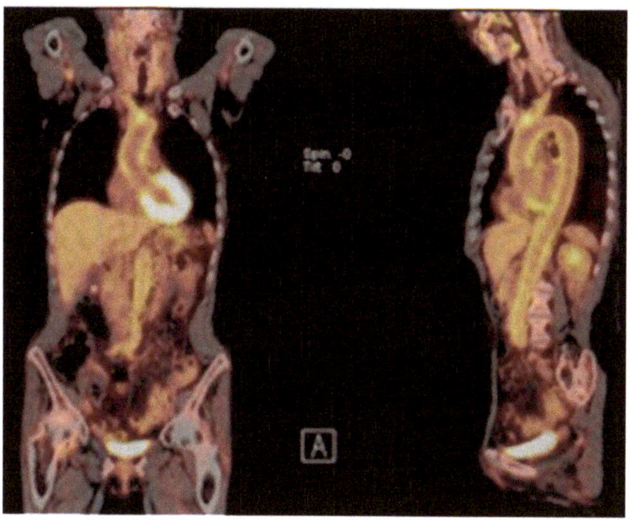

Fig. 2.10 FDG-PET/CT of a patient with a large-vessel vasculitis. The FDG uptake in the aortic wall is compatible with inflammation

2.3.5 Neurology

Several tracers are available for brain imaging [38, 39]. FDG can be used to examine the brain function and is used in dementia disorders. The metabolic activity correlates with brain neuronal function. An interesting compound is the

Pittsburgh substance, useful for imaging of dementia [40]. Other tracers can be used for imaging of the dopaminergic system, ^{11}C-raclopride for imaging of D2-receptors and DOPA to measure the dopamine production in striatum [41]. As for SPECT there is also a PET tracer to measure dopamine reuptake in striatum, ^{11}C-BCIT. The different studies of the dopamine system, pre- and postsynaptic, are mainly used for parkinsonian diseases.

2.4 Further Developments

A rapid development is still on-going, and a lot effort is put on processing of image data. This makes it possible either to decrease the acquisition time or to reduce the administered dose with remained or increased image quality. Acquisition during respiratory gating can give more detailed information for treatment planning if knowing where the tumour moves during the respiratory cycle. The respiratory gated images, 4D imaging, might also be converted into a more correct motion free image by advanced processing and morphing of image data. A goal for the development in imaging is to gain so much biological information about the disease that a personalized treatment can be decided for each patient and that each patient treatment can be individually assessed. Hybrid imaging could contribute to the goal to a large extent but information from other modalities such as MR and other biological markers are warranted to reach the goal. Most of the information is complementary not competing, and deeper knowledge of diseases is needed to take advantage of the information "hidden" in imaging.

2.5 Summary

Nuclear medicine is imaging of different physiological pathways in the body. The used radiopharmaceutical or radiotracer decides which function or molecular pathway is imaged. Introduction of PET/CT, mainly with FDG, has changed the handling in oncology substantially and made image-based personalized medicine in oncology possible [42].

References

1. Hertz S, Roberts A, Evans RD (1938) Radioactive iodine as an indicator in the study of thyroid physiology. Exp Biol Med 38(4):510–513
2. Lin E, Alessio A (2009) What are the basic concepts of temporal, contrast, and spatial resolution in cardiac CT? J Cardiovasc Comput Tomogr 3(6):403–408
3. Ritman EL (2008) Vision 20/20: increased image resolution versus reduced radiation exposure. Med Phys 35(6):2502–2512

4. Lecomte R (2009) Novel detector technology for clinical PET. Eur J Nucl Med Mol Imaging 36(Suppl 1):S69–S85
5. Poeppel TD, Krause BJ, Heusner TA, Boy C, Bockisch A, Antoch G (2009) PET/CT for the staging and follow-up of patients with malignancies. Eur J Radiol 70(3):382–392
6. Cook GJ (2010) PET and PET/CT imaging of skeletal metastases. Cancer Imaging 10:1–8
7. Ben-Haim S, Israel O (2009) Breast cancer: role of SPECT and PET in imaging bone metastases. Semin Nucl Med 39(6):408–415
8. Beheshti M, Langsteger W, Fogelman I (2009) Prostate cancer: role of SPECT and PET in imaging bone metastases. Semin Nucl Med 39(6):396–407
9. Weih M, Degirmenci U, Kreil S, Lewczuk P, Schmidt D, Kornhuber J et al (2010) Perfusion imaging with SPECT in the era of pathophysiology-based biomarkers for Alzheimer's disease. Int J Alzheimers Dis 2010:109618
10. Farid K, Caillat-Vigneron N, Sibon I (2011) Is brain SPECT useful in degenerative dementia diagnosis? J Comput Assist Tomogr 35(1):1–3
11. Gerasimou GP, Aggelopoulou TC, Costa DC, Gotzamani-Psarrakou A (2006) Molecular imaging (SPECT and PET) in the evaluation of patients with movement disorders. Nucl Med Rev Cent East Eur 9(2):147–153
12. Marshall V, Grosset D (2003) Role of dopamine transporter imaging in routine clinical practice. Mov Disord 18(12):1415–1423
13. Aarsland D, Kurz M, Beyer M, Bronnick K, Piepenstock NS, Ballard C (2008) Early discriminatory diagnosis of dementia with Lewy bodies. The emerging role of CSF and imaging biomarkers. Dement Geriatr Cogn Disord 25(3):195–205
14. Roach PJ, Bailey DL, Harris BE (2008) Enhancing lung scintigraphy with single-photon emission computed tomography. Semin Nucl Med 38(6):441–449
15. Reid JH, Coche EE, Inoue T, Kim EE, Dondi M, Watanabe N et al (2009) Is the lung scan alive and well? Facts and controversies in defining the role of lung scintigraphy for the diagnosis of pulmonary embolism in the era of MDCT. Eur J Nucl Med Mol Imaging 36(3):505–521
16. Kluetz PG, Meltzer CC, Villemagne VL, Kinahan PE, Chander S, Martinelli MA et al (2000) Combined PET/CT imaging in oncology. Impact on patient management. Clin Positron Imaging 3(6):223–230
17. Akhurst T, Chisin R (2000) Hybrid PET/CT machines: optimized PET machines for the new millennium? J Nucl Med 41(5):961–963
18. Sweet WH, Brownell GL (1955) Localization of intracranial lesions by scanning with positron-emitting arsenic. J Am Med Assoc 157(14):1183–1188
19. Tewson TJ, Welch MJ, Raichle ME (1978) [^{18}F]-Labeled 3-deoxy-3-fluoro-D-glucose: synthesis and preliminary biodistribution data. J Nucl Med 19(12):1339–1345
20. Reske SN, Kotzerke J (2001) FDG-PET for clinical use. Results of the 3rd German Interdisciplinary Consensus Conference, "Onko-PET III", 21 July and 19 September 2000. Eur J Nucl Med 28(11):1707–1723
21. Hillner BE, Siegel BA, Shields AF, Liu D, Gareen IF, Hunt E et al (2008) Relationship between cancer type and impact of PET and PET/CT on intended management: findings of the national oncologic PET registry. J Nucl Med 49(12):1928–1935
22. Hillner BE, Siegel BA, Liu D, Shields AF, Gareen IF, Hanna L et al (2008) Impact of positron emission tomography/computed tomography and positron emission tomography (PET) alone on expected management of patients with cancer: initial results from the national oncologic PET registry. J Clin Oncol 26(13):2155–2161
23. Eradat J, Abtin F, Gutierrez A, Suh R (2011) Evaluation of treatment response after nonoperative therapy for early-stage non-small cell lung carcinoma. Cancer J 17(1):38–48
24. Mac Manus MP (2010) Use of PET/CT for staging and radiation therapy planning in patients with non-small cell lung cancer. Q J Nucl Med Mol Imaging 54(5):510–520
25. Almeida FA, Uzbeck M, Ost D (2010) Initial evaluation of the nonsmall cell lung cancer patient: diagnosis and staging. Curr Opin Pulm Med 16(4):307–314

26. Gupta T, Beriwal S (2010) PET/CT-guided radiation therapy planning: from present to the future. Indian J Cancer 47(2):126–133
27. Juweid ME (2011) FDG-PET/CT in lymphoma. Methods Mol Biol 727:1–19
28. Cheson BD (2007) The international harmonization project for response criteria in lymphoma clinical trials. Hematol Oncol Clin North Am 21(5):841–854
29. Brepoels L, Stroobants S, De WW, Spaepen K, Vandenberghe P, Thomas J et al (2007) Hodgkin lymphoma: response assessment by revised international workshop criteria. Leuk Lymphoma 48(8):1539–1547
30. Cashen AF, Dehdashti F, Luo J, Homb A, Siegel BA, Bartlett NL (2011) ^{18}F-FDG PET/CT for early response assessment in diffuse large B-cell lymphoma: poor predictive value of international harmonization project interpretation. J Nucl Med 52(3):386–392
31. Cheson BD, Pfistner B, Juweid ME, Gascoyne RD, Specht L, Horning SJ et al (2007) Revised response criteria for malignant lymphoma. J Clin Oncol 25(5):579–586
32. Juweid ME, Stroobants S, Hoekstra OS, Mottaghy FM, Dietlein M, Guermazi A et al (2007) Use of positron emission tomography for response assessment of lymphoma: consensus of the imaging subcommittee of international harmonization project in lymphoma. J Clin Oncol 25(5):571–578
33. Blockmans D, Bley T, Schmidt W (2009) Imaging for large-vessel vasculitis. Curr Opin Rheumatol 21(1):19–28
34. Papathanasiou ND, Du Y, Menezes LJ, Al-Muhaideb A, Shastry M, Beynon H et al (2012) ^{18}F-Fluorodeoxyglucose PET/CT in the evaluation of large-vessel vasculitis: diagnostic performance and correlation with clinical and laboratory parameters. Br J Radiol 85:e188–e194
35. Nishiyama Y, Yamamoto Y, Dobashi H, Kameda T (2010) Clinical value of ^{18}F-fluorodeoxyglucose positron emission tomography in patients with connective tissue disease. Jpn J Radiol 28(6):405–413
36. Bruggink JL, Glaudemans AW, Saleem BR, Meerwaldt R, Alkefaji H, Prins TR et al (2010) Accuracy of FDG-PET-CT in the diagnostic work-up of vascular prosthetic graft infection. Eur J Vasc Endovasc Surg 40(3):348–354
37. Keidar Z, Nitecki S (2009) FDG-PET for the detection of infected vascular grafts. Q J Nucl Med Mol Imaging 53(1):35–40
38. Noble JM, Scarmeas N (2009) Application of pet imaging to diagnosis of Alzheimer's disease and mild cognitive impairment. Int Rev Neurobiol 84:133–149
39. Herholz K, Carter SF, Jones M (2007) Positron emission tomography imaging in dementia. Br J Radiol 80(Spec No 2):S160–S167
40. Nordberg A (2008) Amyloid plaque imaging in vivo: current achievement and future prospects. Eur J Nucl Med Mol Imaging 35(Suppl 1):S46–S50
41. Elsinga PH, Hatano K, Ishiwata K (2006) PET tracers for imaging of the dopaminergic system. Curr Med Chem 13(18):2139–2153
42. Langer A (2010) A systematic review of PET and PET/CT in oncology: a way to personalize cancer treatment in a cost-effective manner? BMC Health Serv Res 10:283

Chapter 3
Physics of Imaging in Nuclear Medicine

Andrej Studen

3.1 Introduction

Imaging in nuclear medicine has become an established part of standard medical practice for recognition and treatment of a wide variety of medical disorders. This chapter gives a compact overview of the physics that governs such imaging algorithms.

All imaging methods in nuclear medicine are based on radioactive nuclei, attached to molecules of interest (radiopharmaceuticals). A specific set of selection criteria has been assembled that identifies such radiopharmaceuticals. The detectors that are used are specifically tailored to capture the emitted radiation. They fall into three categories: scintillators, semiconductors, and light detectors. Based on the examination, the detectors are assembled to diagnostic tools. In single-photon emission computed tomography (SPECT), each photon is detected and processed independently. A set of complex collimators can be used to determine the incoming direction of the impact photon. In positron emission tomography (PET), a pair of photons is both emitted and captured for each valid event. Since the photons in PET are emitted in opposite direction, no mechanical collimation is needed. The line that connects the interactions of both photons is called the line of response. A new variety of a PET device, a time-of-flight (TOF) PET, has such accurate chronometers that the position of emission within the line of response can be determined based on the time difference of both interactions. The image reconstruction is the final step of the imaging algorithm, transforming the results of the measurements into a three-dimensional distribution of radiotracer density.

It is obvious that it is impossible to give all complexity of the field in the allotted space. The reader is kindly referred to the references for further reading and in-depth understanding.

A. Studen (✉)
Jožef Stefan Institute, Jamova 39, 1000 Ljubljana, Slovenia
e-mail: andrej.studen@ijs.si

A. Giussani and C. Hoeschen (eds.), *Imaging in Nuclear Medicine*,
DOI 10.1007/978-3-642-31415-5_3, © Springer-Verlag Berlin Heidelberg 2013

3.2 Radiation Used in Imaging in Nuclear Medicine

In nuclear medicine, radioactivity is used to image metabolic processes characteristic to certain medical conditions. The images provide insight to the function of the tissue, which makes them a functional imaging modality, giving complementary information to anatomical imaging such as CT or plain MRI. The radioactivity is emitted by unstable nuclei chemically bound to a pharmaceutical specific to the imaged process. The isotopes that are used in such procedures are called radionuclides, and compound is called a radiopharmaceutical.

When preparing radiopharmaceuticals [1], the following points have to be considered:

- *Radiation type*. Gamma radiation with energy between 140 and 511 keV is ideally suited for medical applications, since a substantial part (up to 40 % for certain nuclides) passes through body unmodified, and is relatively simple to detect. Radiation of photons with lower energies or other radiation types (alpha and beta particles) is predominantly absorbed in the body. Usage of more energetic photons is rare, since they present a challenge in detection.
- *Half-life*. A physical quantity A, activity, states number of disintegrations of a radioactive sample per unit time. The activity drops exponentially with a nuclide specific half-life $t_{1/2}$:

$$A(t) = A_0 \cdot \mathrm{e}^{-\frac{\ln(2)}{t_{1/2}}t} = A_0 \cdot 2^{-\frac{t}{t_{1/2}}}, \tag{3.1}$$

  where A_0 is the initial activity. The isotopes are chosen so that their half-lives reflect the metabolic processes that are being investigated; in any case, half-lives between hours and days at most are suitable. Table 3.1 lists half-lives of radionuclides commonly used in nuclear medicine.

- *Specific activity*. The specific activity is the activity of the sample divided by its mass. For minimum interference, the specific activity injected should be as high as possible. Although specific activity of *radionuclides* can be very high ($\approx$GBq/pmol), the specific activity of *radiopharmaceuticals* is much lower due to chemical contamination of sample with nonradioactive isotopes of the radionuclide.
- *Purity*. Purity is the portion of activity sourced by the desired compound. Purity should be high, since the remaining radioactivity will (a) expose patients to redundant radioactive dose and (b) add noise to detection. Notorious example is a well-controlled contamination of ^{99m}Tc with its parent ^{99}Mo during extraction from generator.
- *Chemical properties*. The radiopharmaceuticals must mimic all relevant properties of the equivalent pharmaceutical, and elaborate procedures were established to this end. The common methods are a *direct replacement*, where radioactive isotopes are used in place of their stable counterparts, like ^{11}C for ^{12}C in ^{11}C-choline; creation of *analogs*, most notably ^{18}F-FDG as a glucose

Table 3.1 List of common radionuclides for imaging in nuclear medicine

Nuclide	Mode	Photon energy (keV)	Half-life	Imaging modality
^{11}C	β^+	511	20.37 min	PET
^{13}N			9.97 min	
^{15}O			2.04 min	
^{18}F			109.7 min	
^{82}Rb			76.4 s	
^{67}Ga	EC	93, 185, 300	3.26 days	SPECT
^{99m}Tc	IT	140	6.01 h	
^{111}In	EC	172, 247	2.81 days	
^{123}I	EC	159	13.2 h	
^{125}I	EC	27–30	59.4 days	
^{201}Tl	EC	68–80	3.04 days	

For each nuclide, the decay mode (IT is isomeric transition, EC is electron capture, β^+ the positive beta decay), the energy of the photons emitted, the half-life, and the principal imaging modality are given

analog; and *chelation*, e.g., binding of ^{99m}Tc in non-active parts of a complex substance.

- *Production.* Radioactive sources are produced in *nuclear reactors* through chemical purification of fission products or activation and, on a smaller scale, in *cyclotrons*. For convenience, *generators*, i.e., long-lived parents of interesting nuclides (^{99}Mo for ^{99m}Tc or ^{82}Sr for ^{82}Rb) are shipped from remote nuclear reactors to medical facilities. Conversely, cyclotrons are commonly installed at the facilities themselves to provide on-site access to short-lived PET radionuclides (Table 3.1).

3.3 Detection of Radiation

Detectors for nuclear medicine are optimized for detecting γ radiation in 140–511 keV energy range. In this section we will give an overview of the most common interaction types of photons in this energy range and list sensitive materials used for their detection.

3.3.1 Interactions of Photons with Matter

The passage of photons through matter is marked by discrete, point-like interactions. As a consequence, some photons can pass through obstacles unchanged, and it is only their number that diminishes. For an incident current j_0 of photons on a target of thickness d, the outbound current j will be

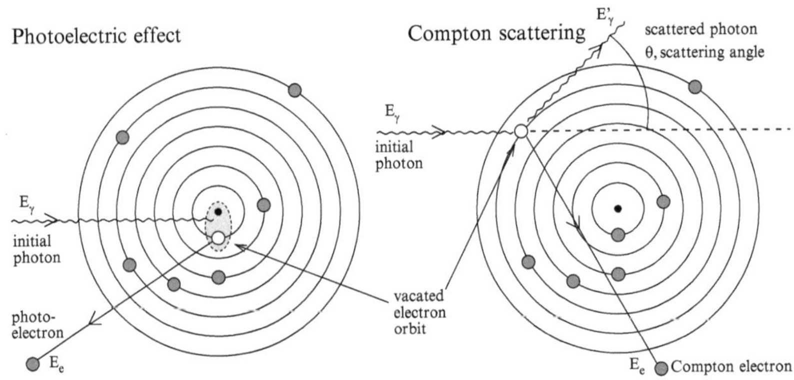

Fig. 3.1 Schematic representation of the most common interactions for photons used in medical imaging: photoelectric effect (*left*) and Compton scattering (*right*)

$$j = j_0 \cdot e^{-\mu \cdot d}, \tag{3.2}$$

where μ is the material specific attenuation coefficient.

Figure 3.1 illustrates the interactions that dominate the specified energy range: the Compton scattering (CS) and the photoelectric effect (PE), each contributing a partial attenuation coefficient to the total attenuation coefficient $\mu = \mu_C + \mu_{pe}$.

In the *photoelectric effect*, the photons are completely absorbed in the electric field of the target nucleus (see Fig. 3.1, left). The energy of the photon is transferred to an electron from the atomic shell of the target atom. The emitted electron is referred to as the photoelectron, and its kinetic energy is equal to the energy of the incident photon minus its binding energy. Since the nucleus is involved in the interaction, probability for PE is higher for electrons in strongly bound shells close to the nucleus. When the energy of the incoming photons exceeds the binding energy of those shells, a sharp rise in the partial attenuation coefficient μ_{pe} can be observed in Fig. 3.2. The most energetic of those edges corresponds to the K-shell transitions, where electrons from the innermost shells (1 s) of the target atoms are excited. Above the K threshold, the attenuation coefficient scales as

$$\mu_{pe} \propto \frac{\rho \cdot Z}{A} \frac{Z^{4-5}}{E^{2.5-3.5}} \tag{3.3}$$

with ρ, A, and Z being the density, mass, and atomic numbers of the target material and E the energy of the incoming photon. The exponents for Z and E vary within the energy range of photons used in imaging. Note the strong Z dependence and the sharp drop with increasing energy of the photons.

The *Compton scattering* is an elastic scattering of a photon with an electron in the atomic shell. The energy transfer between the particles is sufficient to excite the electron, called a Compton electron, from the atom. Since other particles, including

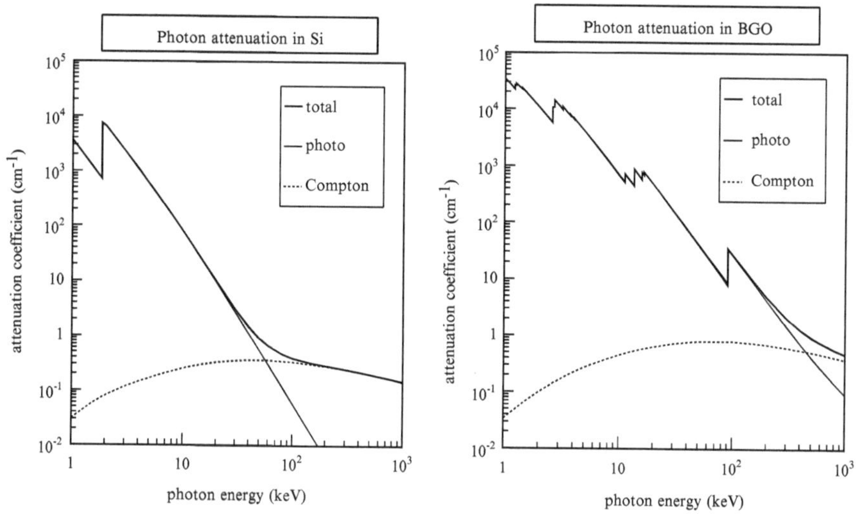

Fig. 3.2 Attenuation coefficient for photons in Si (*left*) and BGO (*right*). Note the logarithmic scales

the nucleus, can be considered as spectators only, a relation between the scattering angle θ of the photon and the transferred energy E_e can be derived:

$$2\,\sin^2\frac{\theta}{2} = \frac{m_e\,c^2\,E_e}{E\cdot(E-E_e)}. \tag{3.4}$$

The quantum effects do not extend to the scattering angle, which is a continuous variable. As a consequence, the spectrum of the emitted electrons is a continuum with a sharp drop, the Compton edge, corresponding to the maximum energy transfer at backscattering. The partial attenuation coefficient μ_C is proportional to

$$\mu_C \propto \frac{\rho\cdot Z}{A} \tag{3.5}$$

and has, conversely to μ_{pe}, a weak relationship to Z and little variation with energy over the given range.

Figure 3.2 illustrates both partial and total attenuation coefficients in two materials, one with a low Z (Si, left) and one with a high Z (BGO, right), which span the range of sensors and materials used in nuclear medicine detection. The strong PE energy dependence breaks the photon interactions to two regions. By labeling E_q as the energy where $\mu_C(E_q) = \mu_{pe}(E_q)$, the photons with energy below E_q will be predominantly photoelectrically absorbed, while above E_q, they will undergo CS, with a narrow region where both processes are equally probable. Below E_q, because of the Z dependence of μ_{pe}, there will be huge difference in attenuation coefficient in different media: at 30 keV for Si with $\mu = 3\ \text{cm}^{-1}$ and

BGO with $\mu = 170\ \text{cm}^{-1}$ the difference is almost two orders of magnitude. Above E_q, μ predominantly scales with ρ, since $Z/A \approx 1/2$ for most materials, resulting in variation confined to less than one order of magnitude.

3.3.2 Sensitive Materials: Scintillators

Scintillators are materials that convert ionization to visual light [2]. For detection of gamma radiation, the signal is produced in two stages—the incident photon undergoes one of the interactions mentioned above to create an electron with substantial kinetic energy, be it Compton- or photoelectron, and then the electron's energy is converted to countable photons of visual light. Table 3.2 lists some properties of common scintillators used in nuclear medicine. The most commonly used scintillator material is NaI, being relatively inexpensive and malleable. For PET, detection efficiency is crucial, and devices are based on BGO (chemically $Bi_4Ge_3O_{12}$) or L(Y)SO. The name L(Y)SO describes two distinct crystals—proprietary LSO (Lu_2SiO_5), and an open source LYSO, which is a 10 % admixture of YSO (Y_2SiO_5). Both have nearly identical properties and will be treated as a single substance throughout the text.

The important parameters of scintillators are (see Table 3.2):

- *Photon yield*. Number of visual photons created per keV kinetic energy of the electron. Higher numbers indicate larger signals enabling better energy and timing resolution. Note advantage of recent engineered materials like L(Y)SO or $LaBr_3$ over BGO.
- *Detection efficiency*. The efficiency of a material can be expressed through the half-value thickness ($d_{1/2} = \ln(2)/\mu$), which is the thickness of material required to detect half of the incoming photon flux. It should be considered that only photons interacting through PE will contribute to the image (Sect. 3.4.3); therefore, it is desirable that the relative Compton scattering probability is small compared to photoelectric effect. Table 3.2 states $d_{1/2}$ and the ratio (CS/PE) for two most common photon energies, 140 keV of ^{99m}Tc and 511 keV of annihilation photons. To maximize efficiency, materials composed of high Z elements ($_{83}$Bi, $_{71}$Lu and $_{53}$I) are used, which benefit from the strong Z^{4-5} dependence of μ_{pe}. Since μ is proportional to ρ, the materials are also relatively dense, from 3.67 g/cm^3 of NaI to 7.3 g/cm^3 for L(Y)SO. At the low end of the energy spectra, almost all materials have sufficiently low values of $d_{1/2}$ and (CS/PE), so that the other properties become crucial for material selection. At the high end, the efficiency becomes crucial; however, the reign of BGO has been overthrown by the superior light yield and lightningly fast decay time of L(Y)SO.
- *Decay time*. Average duration of light emission after the interaction. A short decay time implies large prompt signals and improves timing resolution. Note the excellent values for L(Y)SO and $LaBr_3$.

Table 3.2 Properties of common scintillators in nuclear medicine [3], for $LaBr_3$ see [4]

Material	Yield (ph/keV)	140 keV		511 keV		Decay time (ns)
		$d_{1/2}$ (mm)	CS/PE	$d_{1/2}$ (mm)	CS/PE	
NaI	38	2.83	0.18	21.46	4.6	230
BGO	8	0.59	0.06	7.65	1.3	300
L(Y)SO	32	0.75	0.09	8.55	1.9	40
CsI	52	1.92	0.15	16.93	3.5	1,000
BaF_2	10	2.01	0.18	15.84	4.1	0.6/620
$LaBr_3$	63	2.61	0.26	16.11	6.3	16

For each material, yield, defined as the number of optical photons per keV energy loss of the initial particle (electron!), both half-value thickness ($d_{1/2}$) and ratio of probabilities for Compton scattering versus photoelectric effect (CS/PE) at two characteristic gamma-ray energies (140 and 511 keV) and the decay time of the signal are given

Hardly any material is perfect, and the final choice is governed by the requirements of each specific application.

3.3.3 *Light Detectors*

Scintillators require extremely sensitive photodetectors able to detect single photons. The most common devices are *photomultiplier tubes* (PMTs), composed of a photocathode and a series of multiplication dynodes (8–12 stages are common). On the photocathode, the visual photons are converted (back) to photoelectrons with the quantum efficiency (QE) measuring the conversion rate of photons to photoelectrons emitted towards the first dynode. For standard PMTs, QE ranges between 1 and 3 photoelectrons for ten initial photons depending on the photocathode composition. Subsequent dynodes multiply the signal to about 10^6 times the original photoelectron count. Since the electron multiplication is done in vacuum, the PMTs are in general bulky and fragile and very sensitive to all external influences, even the one of the terrestrial magnetic fields.

Silicon-based light detectors were introduced to imaging in nuclear medicine to circumvent some of those difficulties. Figure 3.3 illustrates the principle of operation. The device is a reversely operated diode with graded dopant concentration ($n^+p\pi p^+$). The uneven doping concentration is reflected in the nonhomogeneous electric field, which is strong in the multiplication region and moderate in the active region. Light enters through the bottom entrance window and produces charge carriers in the active region of the device. A moderate field splits the carriers; the electrons are steered towards the high field region and multiplied. Since charge is multiplied within the crystal, no evacuated container is needed and ordinary magnetic fields will not break the multiplication chain. The same charge containment argument makes QE of silicon-based devices larger than the one of PMTs with a common value of 4 or more carrier pairs per 10 visual photons.

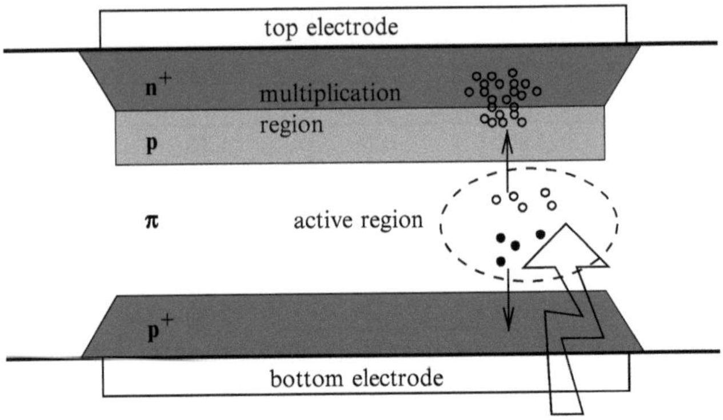

Fig. 3.3 Schematic representation of APD operation

Based on the multiplication factor, two types of devices are constructed:

- In *avalanche photodiodes* or APDs [5], the multiplication factors are between 100 and 1,000, preserving the proportionality of the collected electron current to the incoming light flux. This multiplication factor is at least an order of magnitude smaller than in PMTs, which is normally not a problem, but its strong dependence on temperature, applied voltage, and device nonuniformity is. The timing resolution depends on the multiplication and is also slightly worse than in PMTs.
- The *silicon photomultipliers* or SiPMs [6] operate in Geiger mode, each photon interaction generating a complete device discharge. To keep proportionality, the device is split into tiny cells, squares with sides between 25 and 100 μm. The proportionality is broken when multiple photons interact in a single cell and get counted as a single photon. The dynamic range is limited by the number of cells per device; however, the saturation effects are tolerable, since the devices are generally small (1–10 mm^2) to limit their capacitance and to allow decent timing, and are finely segmented to as much as few thousands of cells per device. Present devices are extremely sensitive to changes in operating conditions, most notable the applied reverse voltage and ambient temperature. Furthermore, the small cell size implies complex readout and electronics.

3.3.4 Sensitive Materials: Solid-State Detectors

Solid-state detectors are modified semiconductor devices. Most sensors are reversely operated diodes, segmented to the desired resolution. The reverse bias depletes the sensor of thermally generated carriers. As a photon strikes the detector and creates a Compton or a photoelectron, the energy of the emitted electron is converted into a countable number of electron–hole pairs. The electron–hole pairs are split in the applied reverse bias and collected to respective electrodes.

Table 3.3 Properties of semiconductors being introduced in nuclear medicine: material, yield given as number of carriers per keV energy loss of the initial electrons, half-value length $d_{1/2}$ and ratio of probabilities CS/PE for Compton scattering versus photoelectric effect, both at 140 and 511 keV photon energy and approximate collection times (they depend on operating conditions and material thickness)

		140 keV		511 keV		
Material	Yield (e–h/keV)	$d_{1/2}$ (mm)	CS/PE	$d_{1/2}$ (mm)	CS/PE	Collection time (ns)
Si	278	20.95	15.7	34.59	490	≈100
CdZnTe	216	2.06	0.23	13.74	6.2	≈100

In general, the number of created pairs in semiconductors, given in Table 3.3, supersedes the yield of scintillators (Table 3.2), giving excellent energy resolution. The diodes can be made to almost arbitrarily small sizes, and resolutions down to 1 μm were measured [7]. The solid-state detectors are currently found in applications where energy or spatial resolutions are crucial (e.g., small animal imaging, Compton camera, high precision scanners [8,9]). The relatively long charge collection times and available manufacturing algorithms favor thinner sensors. To match scintillator efficiencies, multiple layers of segmented sensors have to be used, increasing channel count and device complexities.

3.4 Planar Imaging and SPECT

The most popular radiotracer, ^{99m}Tc, emits a single photon with energy of 140 keV per decay. To detect the origin of the registered photon, Anger [10] developed a mechanically collimated camera shown in Fig. 3.4. The present detectors are improved versions of the same principle, with multiple pinholes instead of a single one, and better scintillators and light detectors.

The sensitive part of the camera consists of a flat scintillator block of NaI or similar material (see Table 3.2), measuring roughly 0.5×0.5 m^2 and has a thickness of 1–2 cm, corresponding to approximately ten half-lengths for ^{99m}Tc radiation. One of the flat faces of the crystal is completely covered by a relatively small number (≈50) of circular PMTs arranged in a hexagonal pattern, illustrated in Fig. 3.5. The opposite side is facing the object through a mechanical collimator that limits the direction of the incoming photons to (nearly) perpendicular to the face of the camera.

The camera registers a planar image, a projection of the source distribution. In analogy to photographic procedure such an image is called a scintigraph and the process scintigraphy. Scintigraphs are used directly in bone scintigraphy where planar image determines presence and approximate location of remote diseases.

Multiple views give volumetric information on source distribution, and rotation of a single camera or imaging with multiheaded cameras is common. The procedure

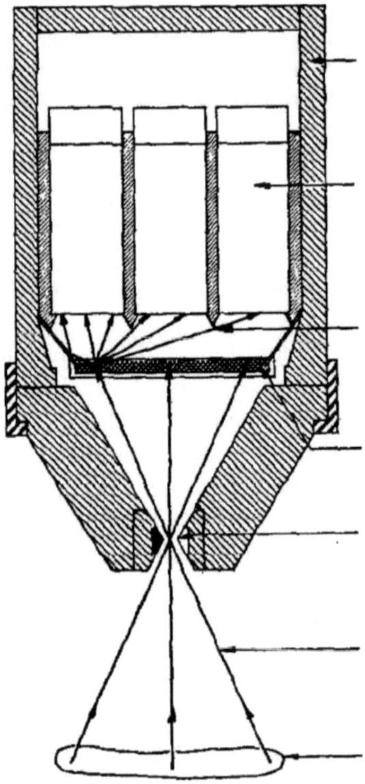

Fig. 3.4 Schematic drawing of Anger camera. From [10]

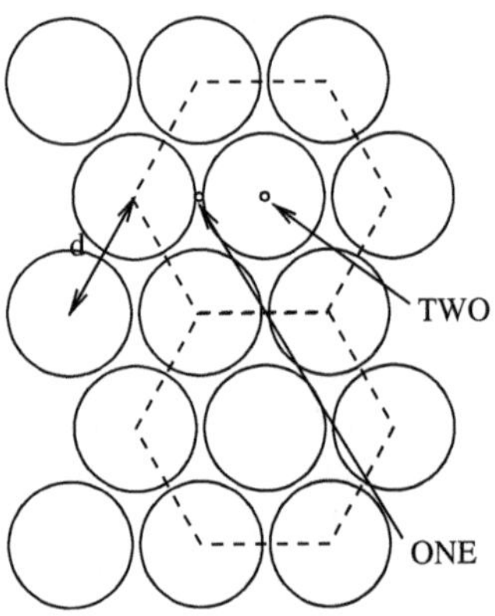

Fig. 3.5 Illustration of hexagonal arrangement of circular PMT on an Anger camera

is called single-photon emission computed tomography (SPECT) since a computer aids in reconstructing the volumetric spatial resolution from detected views. Most frequent examinations with SPECT are studies of myocardial perfusion with a dual-head system.

We will derive basic properties of scintigraphy and SPECT, specifically spatial resolution and efficiency, based on a study of a single planar imaging device.

3.4.1 Spatial Resolution

The spatial sensitivity of the camera comes from the distribution of light among the coupled PMTs. A very common algorithm is the centroid or the center of gravity algorithm. The position where the photon impacts in the camera plane is estimated by the bidimensional quantity **r**:

$$\mathbf{r} = \frac{\sum_{i=1}^{N} s_i \cdot \mathbf{r}_i}{\sum_{i=1}^{N} s_i}, \tag{3.6}$$

where N is the total number of PMTs, $\mathbf{r}_i$ is the projection of the center of the ith PMT on the face of the camera, and s_i is the signal (amount of light) collected in the ith PMT. The position of interaction in direction perpendicular to the camera face is rarely estimated, leading to depth-of-interaction (DOI) artifacts.

This basic strategy is further refined by:

- Omitting PMTs with signals below a certain threshold, since they only contribute noise.
- Aligning gains of individual PMTs.
- Post-correcting with a known response matrix. Prior to imaging, a source is moved in a rectangular grid over the camera face to record distribution of signals among PMTs as a function of source location. During the measurement, the distribution of signals among the PMTs is statistically compared to the collection of pre-calibrated responses. Several minimization techniques are available [11, 12] to spot the most likely candidate, and the position associated with the candidate is the estimated impact position.

The inherent resolution of an Anger camera without a collimator is 3–4 mm FWHM. Surprisingly enough, this resolution is achieved with PMTs as large as 5–10 cm. This size is sufficient because the signal is shared among several PMTs. For that to happen, the scintillator should have a small attenuation coefficient μ for the light it produces. Modern production techniques allow for reasonable regulation of this parameter through crystal growth, edge treatment, and geometry.

The delta method of approximate variance can be used to estimate the variation of the centroid estimator **r**:

$$\mathrm{Var}(\mathbf{r}) = \sum_{i=1}^{N} \left(\frac{\partial \mathbf{r}}{\partial s_i}\right)^2 \mathrm{Var}(s_i). \tag{3.7}$$

The variance of the signal will be dominated by light collection statistics [3], giving $\mathrm{Var}(s_i) = s_i$. The partial derivative can be calculated to yield

$$\frac{\partial \mathbf{r}}{\partial s_i} = \frac{1}{S}(\mathbf{r} - \mathbf{r}_i), \tag{3.8}$$

where S is defined as the total signal: $S = \sum_i s_i$. Since S here is used for illustration of the statistical variation, it makes sense to estimate its value for the weakest link in the signal generation chain, i.e., for the conversion of visual light to photoelectrons on the photocathode. Therefore, in the following, S is the number of created photoelectrons in a given photon interaction.

Should the majority of the signal be collected by M PMTs at an average distance $w = |\mathbf{r} - \mathbf{r}_i|$ from the interaction point, the variance will be just

$$\mathrm{Var}(\mathbf{r}) = \frac{M}{S} w^2. \tag{3.9}$$

The following discussion will be limited to two cases (see Fig. 3.5):

- One, where the interaction point is between two PMTs
- Two, where the interaction point is right inside a given PMT

In the first case, the two nearest tubes will collect most of the signal. Therefore, the distance w corresponds to the radius of the phototube, M is equal to 2, and S to the number of generated photoelectrons in all tubes. For a 140 keV interaction in NaI viewed by a tube with QE of 0.1, the number S is 532 photoelectrons, as can be verified by using the numbers in Table 3.2. Then, even for a $d = 5$ cm diameter tube, the resolution will be

$$\delta_{\mathbf{r}} = 2.35\sqrt{\mathrm{Var}(\mathbf{r})} = 2.35\sqrt{\frac{M}{S}\frac{d^2}{4}} = 3.6 \text{ mm}, \tag{3.10}$$

where a factor of 2.35 was used to convert the square root of the variance to FWHM of the distribution.

In case two, most of the signal is collected in the central PMT, which, with $w = 0$, does not contribute to the variance. All variation comes from the six nearest neighbors which collect a certain portion of the signal. This portion is determined by the product of light attenuation μ and PMT spacing/diameter d:

- If the crystal is opaque, $\mu \cdot d \gg 1$, and there is no signal on the nearest neighbors. As a benefit, there is no additional statistical variation of the position estimate, which always correspond to the center of the tube giving the signal. Unfortunately, the ability to determine position much smaller than the tube dimension is also lost—a 5 cm tube will yield a 5 cm resolution.
- If the crystal is very transparent, there will be signals not only from the nearest neighbors but also from tubes further out. Although the signal is dominated by statistical noise, there are other components to signal variation (electronic noise, drifts in scintillator response, and operating parameters) that will start to accumulate once the contribution of more tubes is added to the signal. The actual

implementation will of course depend on the ability of the system to control this additional noise sources, but most economic way is to allow none but the nearest six tubes to contribute. In this case, M in (3.9) is set to 6, w to tube diameter d, and S to 532, and the variance has to be scaled with the portion of the signal collected on the six nearest tubes (typical values are between 10 and 20 %). The resolution will be

$$\delta_{\mathbf{r}} = 2.35 \times 0.15 \times \sqrt{\frac{M}{S} d^2} = 4.8 \text{ mm}. \tag{3.11}$$

The simplified model correctly accounts for resolution better than tube dimensions but overestimates the positioning errors. Since the actual device resolution is dominated by errors imposed by mechanical collimation, details on positioning algorithms in scintillators can be safely ignored.

3.4.2 Mechanical Collimation

The mechanical collimation is a scheme that allows determination of the direction of the interacting photons based on the known placement of mechanical obstacles placed in photons' path. For example, in Anger camera, all photon paths towards detector save for a hole with diameter $2r$ are blocked by a thick lead block. The most important parameters of the collimator are (illustrated for the Anger camera):

- The spatial resolution, R_{coll}, is the FWHM of the projection of a single point source from the object onto the face of the sensor. For a single circular pinhole, a point source at a distance b will draw a circle with diameter $R_{\text{coll}} = 2 \cdot r \cdot \frac{b+a}{b}$ onto a detector located at a distance a behind the collimator.
- The efficiency, g, is given as a number of photons directed towards the collimator that reach the detector, for a single pinhole $g = \pi r^2 / W$, taking the ratio of solid angle subtended by a circular hole compared to the area of the collimator W.

Mechanical collimators suffer from resolution-efficiency trade-off. For a pinhole, both are related to the hole diameter, and equating for r, the trade-off is expressed as

$$g \propto R_{\text{coll}}^2. \tag{3.12}$$

Relation (3.12) governs mechanical collimation in general, giving rise to multiple collimator species, listed in Table 3.4—there is a high efficiency collimator which sacrifices resolution for efficiency, and there is a high-resolution collimator

Table 3.4 Performance of typical collimators

Collimator type	Gamma energy (keV)	Efficiency	Resolution at 10 cm (mm FWHM)
LEHR	150	1.8×10^{-4}	7.4
LEGP	150	2.7×10^{-4}	9.1
LEHS	150	5.8×10^{-4}	13.1
MEHS	400	1.7×10^{-4}	13.4

LEHR low energy high resolution, *LEGP* low energy general purpose, *LEHS* low energy high sensitivity, and *MEHS* medium energy high sensitivity. The efficiency is given as portion of photons incident on the camera surface that get detected, and resolution is quoted for an object placed 10 cm away from the face of the collimator

which does the opposite. Still, very few of the incident photons are detected even for moderate spatial resolutions, and performance of a mechanically collimated camera is driven by the collimator. Thicker collimators with rougher trade-offs have to be used for more penetrable gamma radiation beyond 140 keV, resulting in a fourfold drop in sensitivity (compare MEHS collimator to LEHS counterpart in Table 3.4).

3.4.3 Energy Resolution and Its Impact on Scattering Removal

Energy resolution of the detector is important in event classification. Consider Fig. 3.6: three types of events can be detected in the sensor. We can deduce position of the source from the last two but not from type 1 events, since in that case the photon directional information was lost due to Compton scattering inside the object. Since this scattering also reduced energy of photon #1, such events are recognized by analyzing the photon energy. A common strategy is to require the energy of the detected photon to be equal to the energy of the initial photon, i.e., selecting type 2 events only. This also disqualifies events of type 3, which by themselves are perfectly legitimate for source position estimation.

The energy resolution of the detector will determine effectiveness of object scattering removal. Consider a photon with energy E undergoing photoelectric effect in scintillator with yield Y coupled to a PMT with quantum efficiency QE. The number of photoelectrons will be

$$S = \mathrm{QE} \cdot Y \cdot E. \tag{3.13}$$

For a typical Anger camera made of NaI and a 140 keV interaction, $S = 532$ photoelectrons with a Poisson type variance $\sigma^2 = S$. The relative energy resolution is $2.35/\sqrt{S} \approx 10\,\%$ FWHM, resulting in a typical admissible energy window of 14 %. A better yield Y gives $\sqrt{Y}$ times better resolution, and a correspondingly narrower energy window can be used to reject scatter and background radiation.

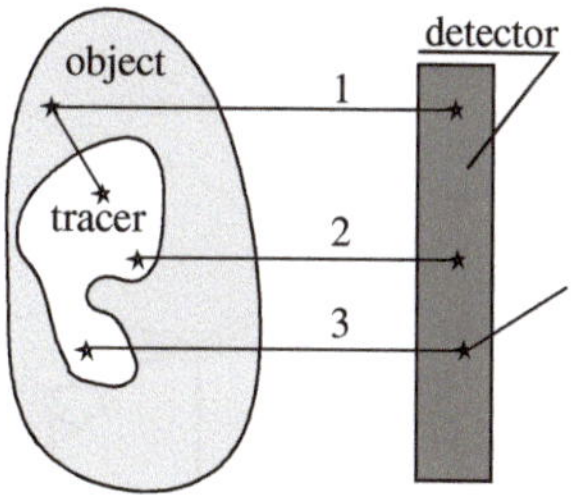

Fig. 3.6 Illustration of different types of events in photon detection

Fig. 3.7 Photograph of a BGO module for a PET

3.5 PET

Positron emission tomography exploits electronic collimation of short-lived positron emitters, like those listed in Table 3.1. The detectors are placed in a stationary ring around the object with a typical ring diameter of $D = 80$ cm and thickness $h \approx 20$ cm. The ring is subdivided into identical detector elements, blocks, which have a high level of autonomy. Figure 3.7 shows a photograph of such a block, consisting of a segmented BGO scintillator block coupled to a set of four PMTs.

The scintillator is segmented to tightly packed bars about 20–30 mm deep along the expected impact direction and with perpendicular dimensions of 4–6 mm. Each block contains between 64 and 144 bars that form a face measuring 5×5 cm^2. The longitudinal cuts between the bars do not reach along their full lengths and stop just before the PMT interface to allow light sharing between neighboring PMTs. A unique signature of signal sharing among PMTs can be recognized for interaction in each of the bars.

Between 100 and 200 blocks are assembled in a full PET ring. Each block provides a trigger signal based on the sum of the signals of all PMTs in the block, with the system trigger based on single triggers from all blocks combined in a complex trigger logic circuit.

The following subsections describe basic properties of PET devices: electronic collimation, spatial resolution, efficiency, and timing resolution (including the principle of TOF-PET). For more details on TOF-PET, see also Sect. 5.4.1.

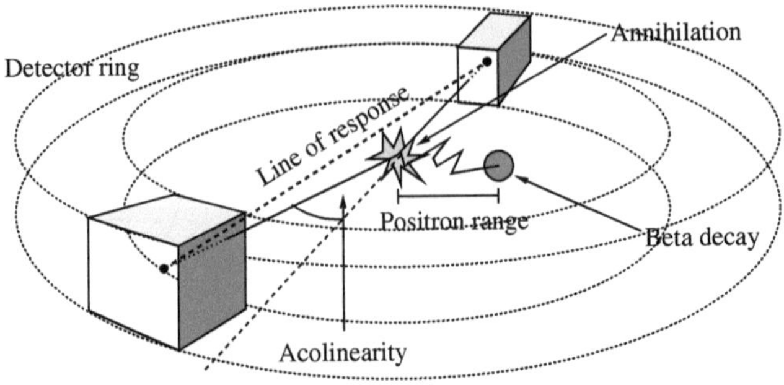

Fig. 3.8 Illustration of PET operation concept

3.5.1 Electronic Collimation

The principle of electronic collimation in PET is illustrated in Fig. 3.8. The radionuclide decays emitting a positron, which annihilates close to the point of emission. The annihilation creates a pair of photons of equal energy (511 keV), traveling in (almost) exact opposite directions. Once both photons are detected, a line of response (LOR) that connects both interactions, and consequently the annihilation point, is constructed. Intersection of many LORs determines the position of the source.

3.5.2 Spatial Resolution

Assuming Gaussian distribution of each parameter, the system spatial resolution R_{sys} is[1]

$$R_{sys} \approx \sqrt{R_{det}^2 + R_{range}^2 + R_{180^\circ}^2} \tag{3.14}$$

with contributions from the detectors (R_{det}), the positron range (R_{range}) and photon acolinearity (R_{180°).

The contribution R_{range}, illustrated in Fig. 3.8, reflects the distribution of annihilation points around the positron emission point. The distribution is cusp shaped with sharp initial drop and long tails. Typical values of FWHM and FWTM (full width at one-tenth of the maximum) of the positron range for common PET

[1] Assuming Gaussian distribution of each parameter. This is most notably not true for the distribution of positron range: however, its contribution is generally small and the resulting deviation irrelevant.

radionuclides range respectively between 0.1 and 1.1 mm and between 1.0 and 4.0 mm, depending on the radionuclide.

Acolinearity derives from the fact that positrons annihilate while still traveling, so photons have to share its momentum and do not, in practice, always fly in exact opposite directions. The error on the estimated annihilation position is $R_{180^\circ} \approx 0.0022 \cdot D$ [13], being D the ring diameter.

In a typical scanner for ^{18}F imaging, the collimation effects R_{range} and R_{180° add up to

$$R_{\text{range}}^2 + R_{180^\circ}^2 = (0.1)^2 + (0.0022 \times 800)^2\,\text{mm}^2 = (1.76\ \text{mm})^2. \tag{3.15}$$

The contribution of the detector resolution to the system resolution is proportional to the resolution of each detector, which is roughly equal to the size of crystal segmentation. Combining R_{range} and R_{180° with bar sizes of 4 or 6 mm, the total resolution R_{sys} equals to 4.37 or 6.25 mm, respectively. For small animal imaging (microPET [14]), bar sizes down to 1 mm are used, which combined with the contribution for acolinearity contribution (scaled due to the smaller ring diameter) provides resolutions of 1 mm FWHM. Even submillimeter resolution was achieved in solid-state detector PET prototypes [15], using the excellent resolution properties of silicon detectors.

3.5.3 *Efficiency*

The detection efficiency g is the portion of emitted photons that contribute to the final image and is a combination of the geometric and the detector efficiencies, reduced by the object absorption.

The geometric efficiency corresponds to the solid angle subtended by the detector at the given annihilation position. The ring arrangement of the PET detector assures full geometric efficiency when both photons are emitted in the plane of the ring, irrespective of the position of annihilation. As soon as photons directed off-plane are included, the efficiency varies with annihilation position and ring geometry. To estimate the scale of efficiency, let us assume an annihilation occurring at the center of the plane that bisects the ring. The angle subtended by the ring is equal for both photons and expressed as

$$P_s = \frac{\Omega_s}{4\pi} = \frac{h \cdot (D/2)}{2 \cdot (D/2)^2} = \frac{h}{D}. \tag{3.16}$$

The efficiency will drop for both off-center and off-plane annihilations.

The detector efficiency is related to the portion of photons incident on the detector that give a valid interaction. For each of the two photons, its detection probability η will be determined by the half-value thickness $d_{1/2}$ of the sensitive material and by the length of the photon path u within the sensor as $\eta = 1 - 2^{-u/d_{1/2}}$ (Table 3.2). Typical scintillator detectors have $u/d_{1/2} \approx 3$ and η of 90 % or more is common. This efficiency is further restricted by the criterion that only photons with energy falling in a selected window are accepted, in order to reject photons scattered in the object. This implicates that also photons undergoing Compton scattering in the detector are rejected. Indeed, ratio of Compton scattering to photoelectric absorption at 511 keV is substantial (see PE/CS, Table 3.2), and consequently, probability of a valid interaction of each of the photons, ε, is equal to η reduced by a factor between 2 (BGO) and 3 (LYSO). Since both photons must give a valid interaction to yield a useful line of response, the total detection efficiency of the ring will equal ε^2.

The portion of emitted photons that are not scattered in the body is given by $Q = e^{-\mu \cdot T}$, where T is the sum of tissue lengths traversed by each of the photon, and μ is the total attenuation coefficient $\mu_{pe} + \mu_C$. Taking a typical distance $T = 10$ cm and $\mu(511 \text{ keV}) = 0.1 \text{ cm}^{-1}$ of water as the main body constituent, $Q \approx 0.36$ is expected.

From the arguments above, we can set the scale of the total detection efficiency g:

$$g = \left(\frac{h}{D}\right)\varepsilon^2 e^{-\mu \cdot T} = 1\,\%. \tag{3.17}$$

The detection efficiency in PET is approximately 2 orders of magnitude larger than in SPECT (compare Table 3.4).

3.5.4 *Timing Resolution in PET, TOF-PET*

The timing resolution of a PET detector corresponds to a variation in delay t_d between arrivals of the trigger signals from detector elements hit by the photons. The timing resolution is important in separating events caused by a single annihilation from an overlap of multiple annihilation events; the latter illustrated in Fig. 3.9, where an obviously erroneous line of response, is assigned to detected photons coming from uncorrelated annihilations.

The synchrony of a pair of interactions in two detector elements is judged by the relative delay of signal arrival from both detector elements. A common filtering method is to select events, where t_d s is within a timing window of duration Δt. The contamination ψ of the sample with random coincidences depends on probabilities of a true, P_{true} or a random event P_{random}, to occur, $\psi = P_{random}/(P_{random} + P_{true})$. In determination of either probability, a valid single interaction starting the timing window stopwatch is assumed, and conditional probability of a second interaction matching the corresponding event type is calculated.

For a true event, the second interaction is caused by the second photon created in the same annihilation as the first interaction. Due to the ring detector geometry,

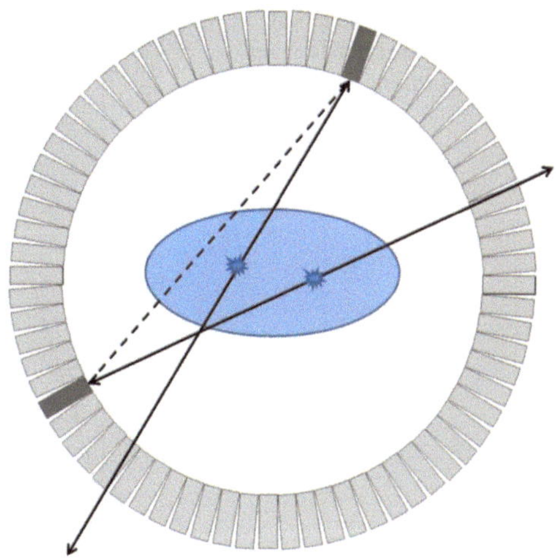

Fig. 3.9 Illustration of improper line of response connecting interactions of subsequent emissions

detection of the first photon sets the second photon on a detector impinging track with a high probability $P_g \approx 1$. Still, effects of limited detector efficiency and energy window, factorized in ε, and object scattering, factorized in Q, must be considered. On top of that, an imperfect timing resolution will further reduce P_{true}. To illustrate the dependence of P_{true} on Δt assume the cumulative delay distribution $P(t_d)$, giving the probability that signal from the second detector module arrived by t_d after the first interaction, is a linear function up to a resolution parameter σ_t and 1 afterwards. In fact, $P(t_d)$ shaped as an error function would be more appropriate, but the general features of such function are implicitly contained in our simplified picture. The shape of P_{true} following that of $P(t_d)$ is shown in Fig. 3.10, saturating at $P_g\varepsilon Q$ for long-timing windows.

For a random event, the second interaction is caused by either of the photons created in the annihilation of the next emitted positron. The interactions in the ring will occur at a rate of $2AP_r\varepsilon Q$, where

- The activity A of the observed object determines the average rate of positron emissions.
- The geometric efficiency P_r is the probability that either of the photons is directed towards the ring, which is, due to the ring geometry, very similar to P_s (3.16).
- Parameters ε and Q are the single-sided detector efficiency and probability of penetration of the photon through the body, respectively.

The waiting time between random interactions in the ring will be exponentially distributed, and probability P_{random} is just the probability that an event will occur within the interval of duration Δt, selected by the first interaction.

$$P_{random} = 1 - e^{-2 \cdot A \cdot P_r \cdot \epsilon \cdot Q \cdot \Delta t}. \tag{3.18}$$

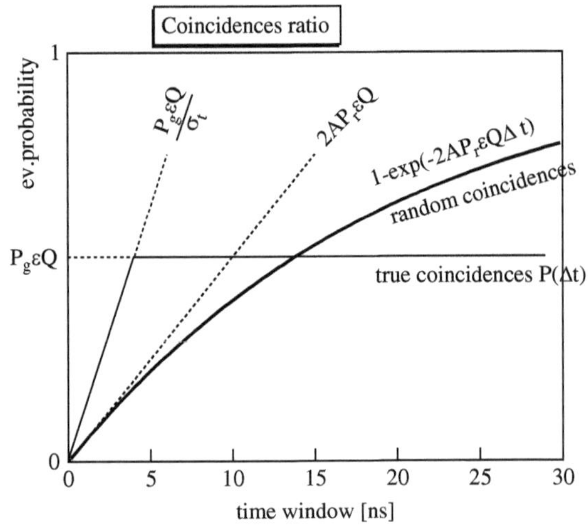

Fig. 3.10 Simple sketch of random and true coincidences ratio versus timing window

Figure 3.10 illustrates the described dependencies. Effect of the filtering is clearly illustrated with a short Δt favoring true events as long as initial slope of P_{true} exceeds that of P_{random}, requiring $\frac{P_g}{\sigma_t} > 2 \cdot A \cdot P_r$. This poses a limit on the required timing resolution for a given activity:

$$\sigma_t < \frac{1}{A} \frac{P_g}{2 \cdot P_r}. \tag{3.19}$$

Roughly speaking, timing resolution should exceed $1/A$, the average time between emissions, with a relief factor $P_g/2{\cdot}P_r$. For the most common case of P_g of almost 1 and P_t on the scale of $h/D = 0.2$ (see Sect. 3.5.3) for a ring geometry, the relief factor amounts to a couple of units. A PET scanner can tolerate activity of as much as 100 MBq sources with a timing windows of 12 ns with a contamination ψ of 10 % or less. The contamination is reduced further by limiting the set of detector elements that are enabled to accept a second interaction to those geometrically opposite to the element signaling the first interaction.

Improving timing resolution beyond that required by (3.19) has other benefits in addition to the reduced contamination ψ. Consider the time-of-flight (TOF) PET principle, illustrated in Fig. 3.11. The photon directed towards left will travel a distance of $D/2 - \Delta x$ before interaction and the right one $D/2 + \Delta x$. This difference translates to a delay of $t_d = 2 \cdot \Delta x / c$ between interactions, with c the speed of light. A detector pair with a resolution of σ_t will be able to separate sources $\delta x = c \cdot \sigma_t / 2$ apart based on timing information. The timing resolution in scintillators scales as [16]

$$\sigma_t \propto \sqrt{\frac{\tau_d}{S}}, \tag{3.20}$$

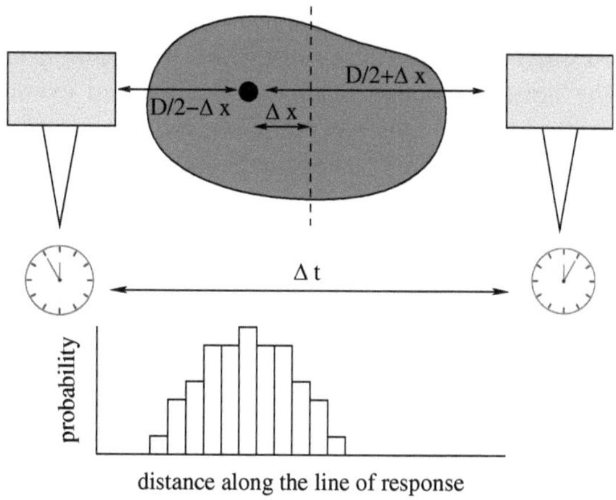

Fig. 3.11 Illustration of time-of-flight PET. Both detectors measure time of interaction, based on the measured delay, a probability profile along the line of response is created, assigning different weights to different positions

where S (3.10) is the photoelectron count. Looking at Table 3.2, we see that LYSO ($Y = 32$, $\tau_d = 40$ ns) will be approximately five times better than BGO ($Y = 8$, $\tau_d = 300$ ns). A resolution of $\sigma_t = 500$ ps, possible in L(Y)SO, gives a $\delta x = 7$ cm. This is actually much worse than the spatial resolution of 4.35 mm estimated above. Therefore, timing information is not used directly in source reconstruction, but it can be useful in noise control. Consider an image reconstructed from non-TOF data: all points along the line of response must be assigned equal probability. In TOF, however, it is possible to assign different probabilities to different segments along the LOR, when t_d is measured (see histogram illustrated in Fig. 3.11). The net effect is as much as twice the information content per detected photon pair in typical TOF-PET system compared to a non-TOF-PET system [17]. This gives either a twice improved image for a fixed event count or, conversely, half the events required for a given image quality.

3.6 Image Reconstruction

Finally, it is the task of image reconstruction techniques to resolve the actual volumetric distribution of the radiation sources based on the collected measurements. Images in a standard sense can be obtained either by projecting or slicing this distribution along a certain object axis with source densities represented by an appropriate selection of shades. The process is rather simple for scintigraphs—they are by themselves projections to detector plane, and images are obtained by segmenting the detector plane to bins and counting interactions in each bin. A true volumetric measurement techniques such as SPECT and PET, requires a complete measurement, that is, a view of the object from all sides. Then the object is segmented to cubes which are called voxels, and, based on the measurements, each voxel gets assigned a certain number of sources. Planar images are formed as slices of this volumetric histogram. The process of obtaining source intensities from measurements is in principle simple but complicated in real life due to small signal-to-noise ratio. Here, the signal is the count of the good events, and the noise is contributed by mis-collimated events, such as unrecognized object scattering, collimator penetration, random coincidences, events from background radiation, etc. This makes image reconstruction a crucial step in detection process and complex reconstruction techniques were developed featuring efficient noise suppression. These exceed the scope of this chapter and will be discussed later in this book.

References

1. Holzwarth U (2011) Radiopharmaceutical production. In: Cantone MC, Hoeschen C (eds) Radiation physics for nuclear medicine. Springer, Berlin, pp 71–103
2. Veronese I (2011) Scintillators and semiconductor detectors. In: Cantone MC, Hoeschen C (eds) Radiation physics for nuclear medicine. Springer, Berlin, pp 161–174
3. Knoll GF (1999) Radiation detection and measurement, 3rd edn. Wiley, New York, NY
4. van Loef EVD, Dorenbos P, van Eijk CWE, Krmer KW, Gdel HU (2002) Scintillation properties of labr3:ce3+ crystals: fast, efficient and high-energy-resolution scintillators. Nucl Instrum Methods Phys Res A 486:254–258
5. Locker RJ, Huth GC (1966) A new ionizing radiation detection concept which employs semiconductor avalanche amplification and the tunnel diode element. Appl Phys Lett 9:227–230
6. Bondarenko G, Buzhan P, Dolgoshein B, Golovin V, Guschin E, Ilyin A, Kaplin V, Karakash A, Klanner R, Pokachalov V, Popova E, Smirnov K (2000) Limited Geiger-mode microcell silicon photodiode: new results. Nucl Instrum Methods Phys Res A 442:187–192
7. Straver J, Toker O, Weilhammer P, Colledani C, Dulinski W, Turchetta R, Bosisio L (1994) One micron resolution with silicon strip detectors. Nucl Instrum Methods Phys Res A 348:485–490
8. Llosa G, Bernabeu J, Burdette D, Chesi E, Clinthorne NH, Honscheid K, Kagan H, Lacasta C, Mikuz M, Modesto P, Rogers WL, Studen A, Weilhammer P (2008) Last results of a first Compton probe demonstrator. IEEE Trans Nucl Sci 55(3):936–941
9. Meier D, Wagenaar DJ, Chen S, Xu J, Yu J, Tsui BMW (2011) A SPECT camera for combined MRI and SPECT for small animals. Nuclear Instrum Methods Phys Res A 652(1):731–734, ISSN 0168-9002, 10.1016/j.nima.2010.09.116
10. Anger HO (1958) Scintillation camera. Rev Sci Instrum 29:27–33
11. Barrett HH, Hunter WCJ, Miller BW, Moore SK, Chen Y, Furenlid LR (2009) Maximum-likelihood methods for processing signals from gamma-ray detectors. IEEE Trans Nucl Sci 56:725–735
12. Tavernier S, Bruyndonckx P, Leonard S, Devroede O (2005) A high-resolution pet detector based on continuous scintillators. Nucl Instrum Methods Phys Res A 537(1–2):321–325. Proceedings of the 7th international conference on inorganic scintillators and their use in scientific and industrial applications
13. Cherry SR, Sorenson JA, Phelps ME (2003) Physics in nuclear medicine, 3rd edn. Saunders, Philadelphia, PA
14. Kim JS, Lee JS, Im KC, Kim SJ, Kim S-Y, Lee DS, Moon DH (2007) Performance measurement of the microPET focus 120 scanner. J Nucl Med 48(9):1527–1535
15. Park S-J, Rogers WL, Huh S, Kagan H, Honscheid K, Burdette D, Chesi E, Lacasta C, Llosá G, Mikuž M, Studen A, Weilhammer P, Clinthorne NH (2007) Performance evaluation of a very

high resolution small animal pet imager using silicon scatter detectors. Phys Med Biol 52(10):2807
16. Choong W-S (2009) The timing resolution of scintillation-detector systems: Monte Carlo analysis. Phys Med Biol 54(21):6495
17. Tabacchini V, Mettivier G, Conti M, Russo P (2010) Improvement in signal-to-noise ratio at variable random fraction in TOF pet. IEEE nuclear science symposium conference record (NSS/MIC)

Chapter 4
Tomographic and Hybrid Imaging in Nuclear Medicine

Axel Martinez-Möller

4.1 Introduction

The first use of a radiotracer for a diagnostic procedure was in 1925 by Hermann Blumgart and Otto Yens, using a modified Wilson cloud chamber to measure the arm-to-arm circulation time [1]. Since then, the instrumentation in nuclear medicine has greatly evolved.

Tomographic and hybrid systems are used in nuclear medicine to provide the three-dimensional distribution of radioactively labeled compounds (tracers) administered to the patient. The tracers are used to study biochemical and physiological processes in vivo based on the tracer principle, whereby radioactive compounds behave in the same way as their nonradioactive counterparts. Using the high sensitivity of nuclear imaging modalities, a small amount of tracer is sufficient to measure these processes without significantly influencing them.

In the first two sections of this chapter, we describe the two types of tomographic imaging systems used in nuclear medicine, Single-photon emission computed tomography (SPECT) and positron emission tomography (PET). In the third section, hybrid systems resulting from the combination of PET or SPECT with CT or MRI are presented.

4.2 Single-Photon Emission Computed Tomography

As discussed in the previous chapter, planar imaging with a gamma camera provides a two-dimensional projection of the spatial distribution of a radiotracer administered to the patient. This measurement can be extended to a tomographic

A. Martinez-Möller (✉)
Department of Nuclear Medicine, Technische Universität München, Munich, Germany

European Patent Office, Grasserstr. 9, 80339, Munich, Germany
e-mail: a.martinezmoller@gmail.com

A. Giussani and C. Hoeschen (eds.), *Imaging in Nuclear Medicine*,
DOI 10.1007/978-3-642-31415-5_4, © Springer-Verlag Berlin Heidelberg 2013

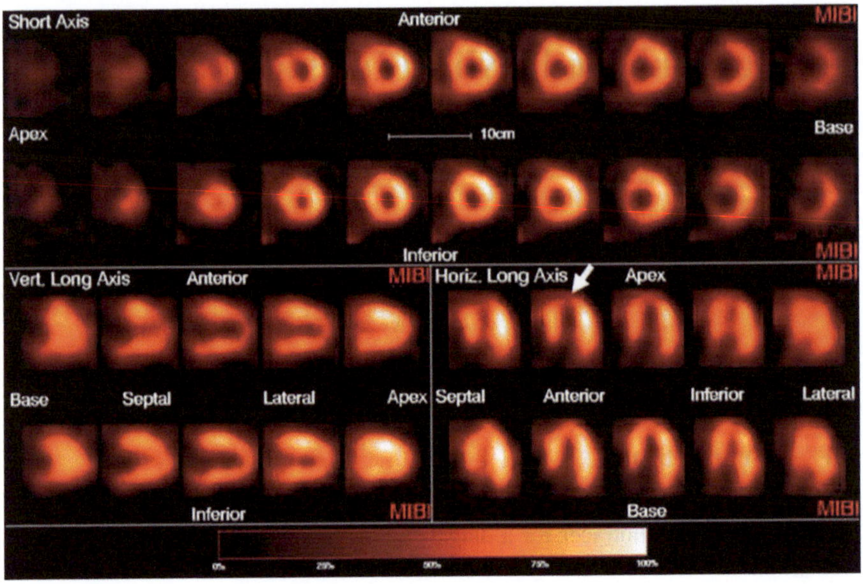

Fig. 4.1 Example of a report page of a rest/stress myocardial perfusion study using ^{99m}Tc-sestamibi SPECT. The report shows the images reoriented according to the cardiac axis (short cardiac axis on the *upper* two rows, *vertical* long axis on the *lower left* rows, *horizontal* long axis on the *lower right* rows). For each view, the stress image is shown on *top* and the rest image below. The *white arrow* indicates an area of reduced perfusion in the apex in the stress image

procedure by acquiring projections from different angles and combining them with image reconstruction techniques to determine the three-dimensional distribution of the radiotracer. This is the basic principle of SPECT, which in its original form consists simply of a gamma camera mounted on a rotation gantry.

4.2.1 Main Clinical Applications

Cardiology accounts for most SPECT examinations. The primary use is for myocardial perfusion imaging [2], allowing identification of hemodynamically relevant coronary stenoses and diagnosis of obstructive coronary artery disease. This is usually achieved by comparing the cardiac images of a flow tracer acquired at rest and at stress (Fig. 4.1), with the stress being induced either by exercise or pharmacologically; tracers such as ^{99m}Tc-sestamibi, ^{99m}Tc-Tetrofosmin, or ^{201}Thallium are most frequently used for this purpose [3]. Myocardial perfusion imaging with SPECT has been reported to have a sensitivity of 86 % and a specificity of 74 % in a pooled analysis of 79 studies including 8,964 patients [4]. A further indication of SPECT in cardiology is the assessment of myocardial viability [5, 6], in order to decide whether a patient can benefit from a revascularization procedure.

Besides cardiology, SPECT is also frequently used in oncology and neurology, as well as imaging of infection and inflammation [7]. Small-animal SPECT systems are also used extensively for preclinical research in these same areas [8, 9].

4.2.2 Principles of the Conventional SPECT Data Acquisition

Although some recent scanners propose innovative architectures and different acquisition procedures, we will describe here the acquisition on SPECT systems based on a rotating gamma camera, which are still the most common ones and serve as a starting point to understand different SPECT architectures.

The conventional SPECT acquisition with a rotating gamma camera is done using the *step-and-shoot* mode: a projection is acquired for some time (shoot), the camera rotates (step), the next projection is acquired, the camera rotates again, etc., acquiring projections at a given number of evenly spaced angular locations (Fig. 4.2).

An important acquisition parameter is the total angle coverage. In principle, projections covering only 180° of rotation are required for reconstruction: opposing views should, when using an ideal projection, contain the same information and thus be redundant. However, due to the spatial-dependent resolution as well as the attenuation and scatter of the photons within the patient, opposing views are not identical to each other in SPECT. Acquiring a 360° acquisition can thus improve the image quality for some imaging tasks and is occasionally preferred [10]. Additional choices of the acquisition are the rotation angle between projections and the acquisition time for each projection which can be defined alternatively as the number of detected photons for each projection. These choices usually reflect a compromise between the targeted image quality and the total acquisition time and vary according to the sensitivity of the scanner.

As described in the previous chapter, the collimation used in a gamma camera is a critical factor for its performance. The same is valid for SPECT: the collimation reflects the compromise between sensitivity and spatial resolution [11]. Most clinical SPECT systems use gamma cameras with parallel-hole collimation; often, a range of parallel-hole collimators is included, among which the user can choose according to the imaging task and the radioisotope involved (see Chap. 3, Table 3.4).

After acquisition, the image reconstruction algorithm and the parameters chosen for the algorithm play a fundamental role in the resulting SPECT images. This topic will be covered in later chapters of this book.

4.2.3 Static, Gated, Dynamic, and Listmode Acquisitions

By recording and sorting the detected photons in different ways, different acquisition modes are possible:

- *Static*. The basic SPECT acquisition is the static acquisition, where the tracer distribution is assumed to be static during the examination. A single image is

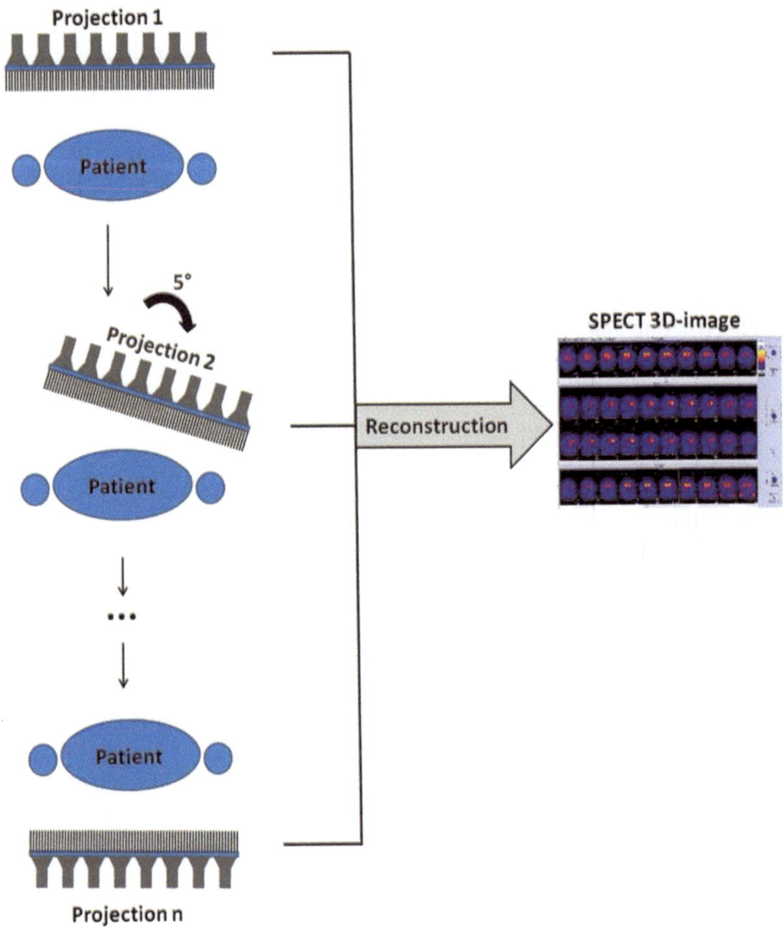

Fig. 4.2 Principle of the conventional SPECT data acquisition with a rotating gamma camera. In this example, projections are acquired with an angular step of 5° over 180°. Using all projections, a three-dimensional SPECT image is reconstructed

generated showing the average tracer distribution over the complete acquisition time.

- *Gated*. In a gated acquisition, the acquired data is sorted into different bins according to the phase at each time-point of a physiological process, typically the cardiac beating or respiration. Subsequently, one image (*gate*) corresponding to each phase can be reconstructed. For cardiac SPECT, it is common to acquire and reconstruct cardiac-gated images [12, 13]: by recording an ECG simultaneously with the SPECT acquisition and sorting the detected photons in different images according to the cardiac phases, a sequence of images (typically 8–16) from systole to diastole is obtained. The image series can be used to analyze the ventricular function and the myocardial wall motion. It is important to note that the image of each cardiac phase is not acquired during a single beat but as the average over several cardiac cycles. In patients with arrhythmia, irregular beats are usually rejected in order to correctly capture the cardiac beating. For cardiac imaging, it is common to perform an ECG-gated acquisition and then compute a static image as the sum of all gates. In some situations such as lung imaging, respiratory gating is applied [14], which is similar to cardiac gating except that the acquisition is triggered with a respiratory signal rather than with the ECG; the resulting image series ranges then from end-inspiration to end-expiration.

- *Dynamic.* The dynamic acquisition mode provides a temporal sequence of images showing the change of tracer distribution with time [15]. The data is sorted here into temporal bins according to a temporal framing scheme predefined by the user. Due to low sensitivity and subsequently long acquisition time, dynamic acquisitions are uncommon with conventional SPECT systems but might gain importance in the future with newer SPECT systems.
- *Listmode.* The most flexible way to acquire the SPECT data is the so-called *listmode* acquisition, in which each detected photon is recorded together with the time of detection. In parallel, physiological signals (ECG or respiration) might be recorded. After acquisition, the user decides how to sort the listmode data to arbitrarily generate static, gated, and/or dynamic images, according to the purpose of the examination. Listmode format thus allows preserving the flexibility to post-process the data in different ways. However, in clinical routine, this flexibility is rarely needed, and the acquisition is seldom recorded in listmode format because of the large storage space required.

4.2.4 Improvements in SPECT Technology

Different technologies have been developed to enhance the conventional SPECT acquisition as described in the previous paragraphs. The most relevant advances are presented below, excluding hybrid SPECT/CT imaging which will be detailed in a later section.

- *Multiheaded cameras.* Multiheaded cameras consist in the simultaneous use of several rotating gamma cameras. They increase the fold of sensitivity by the number of heads, allowing a faster SPECT acquisition with preserved image quality. The most common configuration, two-headed SPECT, has two cameras mounted at, for example, 90° or 180°, of each other. Also triple-headed systems with 120° spacing are available. Multiheaded cameras have been available for more than two decades and represent nowadays the majority of the installed SPECT systems.
- *Continuous and pseudo-continuous acquisition.* In the step-and-shoot acquisition, the camera does not acquire data during each rotation step. By using a continuous acquisition mode with a large number of small rotation steps, the

counting efficiency can be maximized and a higher angular sampling can be achieved [16]. Therefore, some systems allow continuous acquisition modes or, more frequently, pseudo-continuous modes which perform sequential rotational steps but keep acquiring data during the rotational motion.

- *Noncircular acquisition.* In the conventional SPECT architecture, the spatial resolution and sensitivity decrease with increasing distance between the imaged object and the detectors. In order to minimize the distance between the patient and the detectors, noncircular acquisitions can be used. The gamma camera follows elliptical orbits or tracks and contours the patient's body, effectively minimizing the distance and improving the spatial resolution [17, 18].
- *Collimator design.* Since the development of the first gamma camera, different collimator designs were investigated, and their potential advantages were recognized [11]. However, parallel-hole collimators have been used in most systems. Recently, clinical systems which use alternative collimator designs such as slit/slats or multi-pinhole collimation have been commercialized. These have been shown particularly useful when the examination focuses on a small region of interest (e.g., cardiac imaging), resulting in higher sensitivity and spatial resolution for this area. In small-animal systems, submillimeter resolution has been demonstrated with multi-pinhole collimation [19–21].
- *Attenuation correction.* A large fraction of photons interact with the patient's body. As a result of the interaction, the photons do not reach the detectors or change their original directions before being detected, introducing artifacts in the images (see Chap. 3, Fig. 3.6). These artifacts can be minimized by correcting for scatter and attenuation. These corrections require an attenuation map with the radiodensity of each voxel, and two ways have been developed to measure the attenuation map: using a rotating radionuclide source to acquire a transmission scan through the body (as opposed to the emission scan) [22] or using the CT component in hybrid SPECT/CT scanners. However, attenuation correction is challenging in SPECT because the probability of a photon interacting with the patient's body depends on the depth at which the photon was emitted.
- *Iterative image reconstruction.* Over the past years, there has been a shift from analytical reconstructions, mainly filtered backprojection (FBP), to iterative reconstruction algorithms, which can take into account the physics of the acquisition and the imaging system during reconstruction [23]. Iterative algorithms in SPECT include techniques for resolution recovery or noise suppression, in addition to corrections for scatter and attenuation when an attenuation map is available. More information on the different reconstruction algorithms is provided in subsequent chapters of this book.
- *Solid-state detectors.* The initial design of the gamma camera proposed by Hal Anger, consisting of crystal detectors coupled with an array of photomultiplier tubes, has been successfully used for more than 50 years. Thallium-doped sodium iodide detectors, NaI(Tl), have been traditionally used, with the light

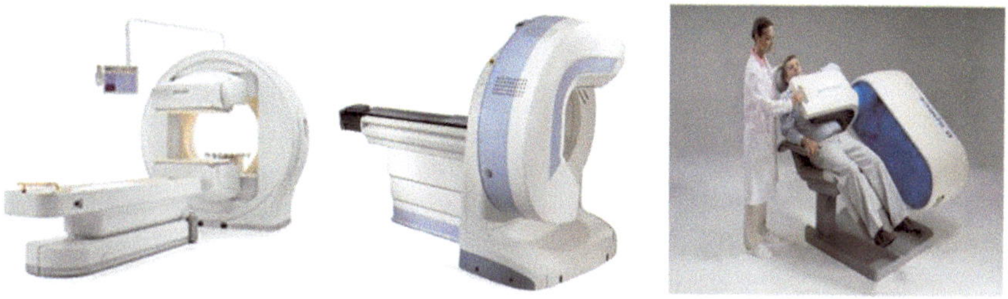

Fig. 4.3 Examples of general purpose SPECT (based on the rotating gamma camera principle) and dedicated cardiac SPECT systems. On the *left*, image of the BrightView (Philips), a general purpose two-headed SPECT camera. The *mid* and *right* figures show two examples of dedicated cardiac SPECT systems with CZT detectors arranged in an arc configuration: the Discovery NM 530c (GE) and the D-SPECT (Spectrum Dynamics)

readout performed with photomultiplier tubes. A major advance in SPECT technology is the introduction of new solid-state detectors, which convert the energy from the incoming photons directly into an electrical signal, replacing both the crystal detectors and the photomultiplier tubes. Solid-state detectors, such as the most commonly used cadmium zinc telluride (CZT), provide a much better energy resolution (about 5 % vs. 10 % at 140 keV for conventional systems) and higher count rate capabilities. As a result, SPECT cameras with solid-state detectors allow better image quality or, alternatively, much shorter acquisition times with preserved image quality [24].

As a result of these advances, there has been a trend in SPECT imaging towards the development of three different subgroups of systems: general purpose SPECT scanners (most frequently dual-headed systems based on the rotating gamma camera design with some enhanced features), hybrid SPECT/CT systems (which will be described below), and dedicated cardiac SPECT systems. The new generation of dedicated cardiac systems changes the standard architecture of a rotating gamma camera, with scanners featuring solid-state technology and static detector sets such as the ones shown in Fig. 4.3. This new generation of dedicated cameras combines a compact design and high photon sensitivity (threefold to tenfold increase as compared to conventional SPECT systems), leading to faster acquisitions while maintaining or increasing the spatial resolution [25–27]. Using these systems, cardiac-gated acquisitions with satisfactory quality can be performed in less than 5 min, as compared to more than 20 min for conventional dual-headed SPECT scanners.

4.3 Positron Emission Tomography

PET systems are designed to measure and quantify the three-dimensional distribution of a positron-emitter using the fact that two collinear 511 keV photons at an angle of $180 \pm 0.25°$ from each other are emitted as a result of the positron

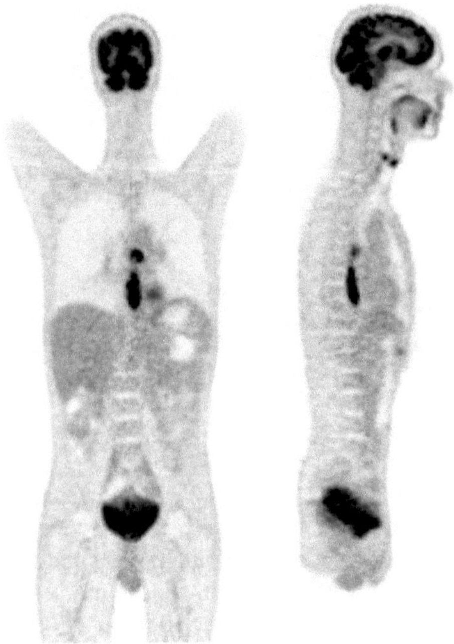

Fig. 4.4 Example of coronal (*left*) and sagittal (*right*) slices of a ^{18}F-FDG PET whole-body image. The image demonstrates increased focal uptake in an esophageal lesion and a mediastinal lymph node. Normal physiological uptake is seen in the brain and bladder

annihilation (see the previous chapter for a detailed description of the underlying physics). Thus, the projections are not made by the detection of single photons as in SPECT but using the detection of pairs of collinear photons arising from the same positron annihilation.

4.3.1 Main Clinical Applications

The most common use of PET is for oncology [28, 29]. Tumor imaging (diagnosis, staging, therapy monitoring, and radiotherapy planning) accounts for over 90 % of all PET examinations. The majority of the examinations use fluorine-labeled fluorodeoxyglucose (^{18}F-FDG) [30–32], a glucose analog showing increased uptake in tumor cells due to their higher glucose metabolism as compared to normal cells (Fig. 4.4) [33]. Besides ^{18}F-FDG, many other PET tracers targeting different biological processes in tumor growth are available [34].

Besides oncology, the other major applications of PET are cardiology [35] and neurology [36], as well as preclinical research with scanners dedicated to small-animal imaging [37]. Other less frequent applications include imaging of gene expression, assessment of infectious disease, and pulmonary function.

4.3.2 Architecture of PET Systems

When two 511 keV photons reach the detectors *nearly* simultaneously (coincidence detection), a positron annihilation is assumed to have occurred somewhere along the line between the two detectors—this line is called *line of response* (LOR) and is actually a parallelepiped when the finite width of the detectors is taken into account. Therefore, no physical collimators are required in PET, but the collimation is done electronically, by considering only pairs of photons reaching the detectors *nearly* simultaneously.

For coincidence detection, at least two opposing sets of detectors must be used. All modern commercial PET scanners (actually hybrid PET/CT scanners, since standalone PET scanners are obsolete) extend this concept by using static rings of detectors placed around the patient. For whole-body scanners the detectors are typically arranged in a gantry of about 70–80 cm diameter.

The detectors are composed of scintillator crystals coupled to photodetectors; the formers convert the incoming photon into light, which is then transformed into an electrical signal. Regarding the scintillators, crystals such as BGO, LSO, LYSO, or GSO are currently used, cut in blocks with a small surface facing the patient (e.g., 4×4 mm^2) and deep enough to detect most 511 keV photons (usually 20-mm depth). For the light readout, photomultiplier tubes are generally used as photodetectors. As the photomultiplier tubes are bulky and small crystals are needed to achieve a high spatial resolution, the coupling between the scintillator and the PMT is not done one-to-one for most clinical scanners but rather in a block design with, for example, a matrix of 8 x 8 crystals coupled to only four photomultiplier tubes. The signal between neighboring PMTs is compared using a coding scheme to determine the scintillator crystal of interaction [38].

Such detectors allow the measurement of the time at which the incoming photon was detected (with the accuracy provided by the time resolution) and the energy which was deposited in the detector (with the accuracy provided by the energy resolution). If the energy deposited is not within a given energy window, for modern clinical scanners usually with a lower threshold of 400–450 keV and a higher threshold of 600–670 keV, the photon is assumed to have undergone Compton scatter and is discarded for further processing. If the energy deposited is within the predefined window, a coincidence window is open: if a second photon with suitable energy is detected within the time defined by the coincidence window, both photons are said to be in coincidence and are considered to arise from the same positron annihilation somewhere along the LOR between the two detectors (see Fig. 4.5a). If the annihilation does not occur at the center of the LOR, the two photons will reach the detectors at a slightly different time; this difference can be as much as 3 ns for annihilations in whole-body scanners (~70 cm diameter with photons moving at the speed of light). Therefore, the coincidence window is typically defined to be about the summation of this time and twice the time resolution of the detectors.

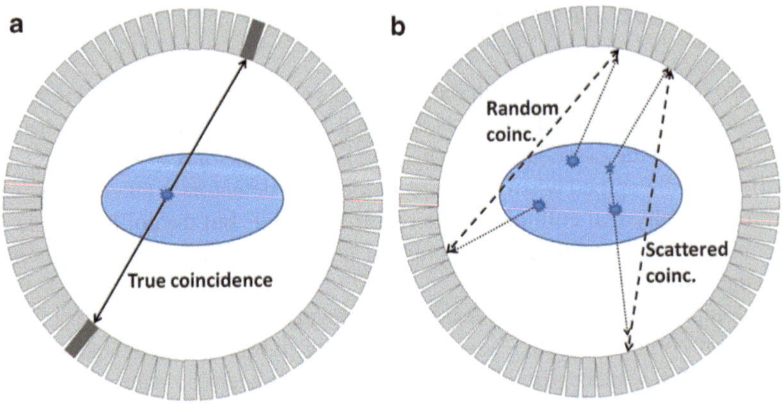

Fig. 4.5 Coincidences in a PET scanner shown in a 2D ring of detectors. On the *left* (**a**), a true coincidence is described: an annihilation event produces two collinear 511 keV photons which move in opposite directions and are detected within a short time window by two detectors; the annihilation event is correctly assigned to be somewhere along the LOR between the two detectors. On the *right* (**b**), random and scatter coincidences lead to incorrectly assigned annihilation events (*dashed lines* corresponding to assumed paths of the photons, *dotted lines* corresponding to real paths of the photons). The random coincidence corresponds to two unrelated annihilation events with only one photon detected for each (because the other photon is absorbed by the patient, is not stopped by the detectors, or goes outside the axial field-of-view of the scanner in 3D). The scattered coincidence corresponds to the detection of two photons originating from the same annihilation, but at least one of them is scattered on its way losing part of its energy but still being accepted by the predefined energy window

The performance of a PET scanner depends strongly on the energy and time resolution of the detectors. If the energy resolution is low, the lower energy threshold must be set to a small value in order to correctly identify most 511 keV photons, but many scattered events will also be accepted (see Fig. 4.5b) [39]. Similarly, two drawbacks are associated with a low time resolution: first, the random detections are increased (Fig. 4.5b), that is, more unrelated photons from different annihilations are considered to be in coincidence degrading the image quality; second, less activity can be present in the field of view (FOV) before photons from different annihilations are detected within the coincidence window (the scanner enters *saturation* earlier). Another advantage of having detectors with high time resolution is the possibility to perform time-of-flight measurements (see Chap. 3, Sect. 3.5.4).

State-of-the-art clinical PET scanners achieve a spatial resolution of about 4 mm, energy resolution of about 11.5 %, and time resolution of about 550 ps. More detailed information about performance evaluation of PET scanners is available in [40].

4.3.3 Principles of PET Data Acquisition

The rings of detectors in clinical PET scanners cover an axial range of 15–22 cm. This is sufficient for single-organ imaging (e.g., brain and cardiac imaging). In oncology, multi-bed acquisitions are usually performed for partial-body or whole-

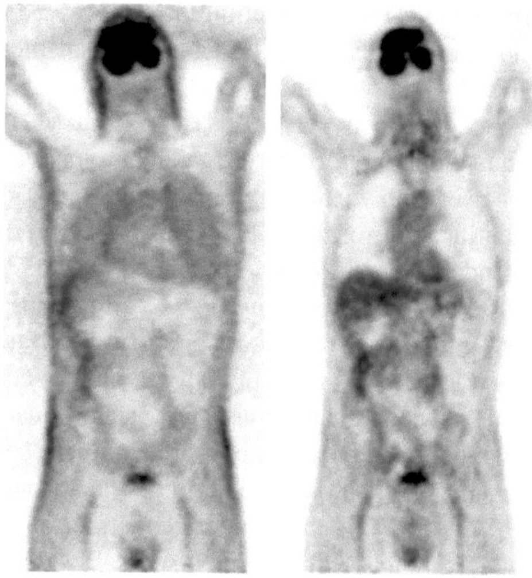

Fig. 4.6 Coronal slice of a ^{18}F-FDG PET whole-body image reconstructed without (*left*) and with (*right*) attenuation correction. The activity distribution in the image without attenuation corrected is distorted, creating an apparent contrast between the skin and the inner structures and a false increased uptake in the lungs

body imaging, following a step-and-shoot acquisition mode. In difference to SPECT, the shoot phase here is the acquisition of a complete 3D image limited only in its axial FOV, and each step is an axial movement of the patient bed. To account for the decrease of sensitivity at the border of the axial FOV, the bed positions are acquired with some overlap; alternatively, continuous bed motion has been suggested [41, 42] but is not yet commercially available. In the multi-bed step-and-shoot acquisition, as usually practiced in whole-body oncology imaging, an acquisition time of 1–4 min per bed position is typically used, depending on the sensitivity of the scanner, the patient weight and the injected dose (taking into account the radioactive decay until the acquisition).

At this point, it is important to understand the role of photon attenuation. Actually, the majority of the 511 keV photons interact with the patient's body, so that only a small fraction is detected–for the inner part of the abdomen, this fraction can be below 5 %. In order to account for this effect and create quantitative PET images, attenuation correction must be performed (Fig. 4.6) [43]. For this purpose, an attenuation map is needed with the *linear attenuation factor* μ at 511 keV for each voxel. The attenuation factor for each LOR is then determined by integrating the attenuation factor over the LOR: the attenuation is independent of the place along the LOR where the annihilation event took place, because the joint trajectory of the two back-to-back photons is the complete path between the detectors.

Therefore, in addition to measuring the photons resulting from the positron annihilations within the patient, an attenuation map is routinely acquired. The acquisition of the attenuation map in standalone PET scanners (now obsolete) was typically done using a rotating radionuclide source, requiring a few additional minutes per bed position. This is not the case anymore for hybrid PET/CT scanners, where the CT image (acquired in a few seconds) is used for this purpose. Whole-body ^{18}F-FDG acquisitions in state-of-the-art PET scanners (hybrid PET/CT) typically last 5–15 min.

As described for SPECT, in PET it is also possible to perform listmode acquisitions and reconstruct gated or dynamic images. In oncology, respiratory gating has been shown to facilitate better tumor delineation and more accurate uptake quantification for thoracic lesions [44]. The disadvantage of this approach is that an extended acquisition time is needed in order to achieve good image quality for each gate. For cardiac imaging, ECG gating is usually performed, and the possibility of performing dual cardiac-respiratory-gated acquisitions has been investigated [45]. Moreover, PET is well suited for dynamic acquisitions because of its high sensitivity. Some physiological parameters can be quantified in absolute terms by analyzing dynamic images with appropriate kinetic modeling. One example is the use of dynamic acquisitions to determine the absolute myocardial blood flow [46].

As in SPECT, the image reconstruction (discussed in a later chapter) has a big impact on the resulting PET images. Besides the algorithm and parameters used, the implementation of different corrections (for attenuation, scatter, randoms, detector normalization, and dead time) also plays an important role.

4.3.4 Improvements in PET Technology

In the last two decades, PET underwent major technological improvements. We list below the most important ones, excluding changes specific to hybrid PET/CT and PET/MR scanners:

- *New scintillators*. From 1980 through the year 2000, bismuth germanate (BGO) was typically used in PET as scintillation crystal. Since then, a new generation of scintillators (including LSO, LYSO, and GSO) demonstrated better performance, combining a high stopping power with decay times much shorter than the ones for BGO. These scintillators were better suited for high count rates and 3D acquisitions.
- *3D acquisition*. Physical collimators (tungsten septa) were previously used in PET to isolate each ring of detectors from the neighbors in order to acquire only in-plane events and avoid detection of oblique photons, referred to as 2D acquisition. Nowadays, PET systems acquire in 3D mode without septa and thus achieve a dramatic increase in sensitivity as compared to 2D acquisition.
- *Smaller crystals*. The spatial resolution of a PET system depends strongly on the size of the scintillator crystals. Most modern PET (PET/CT) systems use crystals with a face of 4×4 mm^2 and depth of 20 mm. Yet, the crystals cannot be made indefinitely small because of the associated loss of volumetric sensitivity.
- *New radiotracers*. Many new radiotracers targeting relevant physiological processes have been developed in PET in recent years. Some of them are already clinically used, and others await authorization for clinical use and might reach the market in coming years. These new radiotracers have a deep impact on the clinical use of PET.

- *Iterative reconstruction.* As in SPECT, there has been a shift from filtered backprojection to iterative reconstruction. Currently, the most commonly used reconstruction in PET is the ordered-subsets expectation maximization algorithm. The implementation of the iterative reconstruction is recently changing from 2D (after data rebinning) to fully 3D reconstruction, supported by the increasing computing power. Resolution recovery techniques are also gaining popularity, which partially compensate the decrease of resolution at the outer part of the FOV.
- *Time-of-flight.* When recording a coincidence event, it is only possible to conclude that an annihilation took place somewhere along the line between the two detectors. If the detectors could resolve the exact arrival time of the photons and by assuming perfect collinearity, it would be possible to precisely localize the location of the annihilation. Recent commercial scanners with fast scintillators and electronics can reach time resolutions below 1 ns, so that the annihilation can be localized down to only a line segment instead of the complete line (e.g., for 500 ps time resolution, the annihilation is located within a 7.5-cm line segment). This technique, known as time-of-flight, results in a significant increase of image quality [47]. With the advent of new photodetectors [48], further improvement of the time resolution is expected. Using digital silicon photomultipliers, it has been shown that a time resolution below 200 ps can be achieved [49].

4.3.5 PET Versus SPECT

PET and SPECT share many similarities in that they both perform functional imaging by means of radioactive probes. However, they also have a number of differences.

- *Isotopes involved.* Although a SPECT scanner can in principle measure the 511 keV photons resulting from a positron-emitter (used in PET), they are optimized for radioisotopes which do not involve positron emission and directly result in the emission of a gamma ray, typically in a lower energy range (e.g., ^{210}Tl at 70 keV, ^{99m}Tc at 140 keV). Positron-emitters used in nuclear imaging have a half-life between 1 min and 2 h and can often be incorporated by replacement into light molecules naturally present in the organism (e.g., $^{15}O–H_2O$). Isotopes clinically used in SPECT usually have a longer half-life (from 6 h for ^{99m}Tc to 3 days for ^{201}Tl or 59 days for ^{125}I) and often can be incorporated only to radiotracers with a high molecular weight which are not naturally present in the organism. The difference in half-lives has a direct impact on their clinical use. PET is ideally suited to study rapidly changing physiological processes because the radiotracers decay quickly. SPECT is better adapted for the study of slower processes and allows imaging hours or even days after tracer administration. Moreover, isotopes with different energies can be simultaneously used and discriminated from each other in SPECT, making dual-isotope imaging possible [50].

- *Sensitivity.* The coincidence detection used in PET allows a much higher sensitivity as compared to the physical collimation used in SPECT, which rejects a large majority of the incoming photons. The sensitivity of PET is about two orders of magnitude that of SPECT; therefore, in PET less radiotracer is often needed to measure a physiological process, and the acquisition can be done in less time.
- *Spatial resolution.* The spatial resolution in PET is limited by two fundamental physical limits: the positron range and the photon acolinearity (see Chap. 3, Sect. 3.5.2). Conversely, the spatial resolution of SPECT is only limited by the resolution of the detectors, so that a very high spatial resolution is theoretically possible. However, the necessary compromise in SPECT between sensitivity and resolution changes this fact, so that in clinical systems the opposite is true: current clinical PET scanners reach a spatial resolution of 4–5 mm, whereas SPECT scanners typically deliver a notably worse resolution of 8–10 mm. High-resolution (<1 mm) SPECT acquisitions at the expense of low sensitivity are sometimes used in small-animal imaging. The resolution of PET in small-animal imaging is about 1.5–2 mm.
- *Attenuation correction.* Correction for photon attenuation is more accurate in PET than in SPECT, where it still represents an exception to perform such a correction. In PET, the detection of two photons at opposing detectors allows an exact attenuation correction: the attenuation factor used for the correction is the integral of the attenuation between the two detectors independent of the location of the annihilation event, because the joint trajectory of the two back-to-back photons is the complete path between the detectors. Attenuation correction is routinely performed in PET and is a critical step to obtain images quantifiable in absolute activity units. In SPECT, it is much more challenging to accurately correct for photon attenuation because photons originated at different depths will suffer different attenuation patterns. There is still controversy about the clinical benefit of attenuation correction in SPECT [51], and most SPECT systems do not yet include mechanisms to measure the tissue attenuation. The lack of attenuation correction leads to artifacts which makes the image interpretation more difficult.
- *Cost.* PET is significantly more expensive than SPECT regarding both the hardware (scanner) and the radiotracers involved. Most radiotracers used in PET require a local cyclotron because of their short half-lives (for some tracers a generator can be used, such as ^{82}Sr, as a generator for ^{82}Rb). In SPECT, tracers are easier to distribute, and the imaging center does not need an expensive cyclotron nearby; thus, the cost of the radiotracers is lower as compared to PET.

The growth of PET in the last years has been much stronger than the growth of SPECT, especially since the advent of hybrid systems. It has been even discussed whether, in the long term, SPECT can progressively decline and become a secondary imaging modality in favor of PET: the higher sensitivity, improved image quality, and the availability of new tracers targeting different physiological processes seem to support this statement [52]. However, the new generation of solid-state SPECT systems shows that SPECT can still be very competitive. The economical cost of switching from SPECT to PET is high, the clinical benefits are not obvious in all imaging tasks, and it seems more likely that both modalities coexist and continue to develop in parallel [53–55].

4.4 Hybrid Systems

4.4.1 Brief History of Hybrid Systems in Nuclear Medicine

In medical imaging, two kinds of modalities can be distinguished: anatomical imaging, which allows an accurate visualization of the morphology (the structure of the organism), and functional imaging, which depicts physiological processes (fast dynamic changes within the organism). The benefit of correlating functional imaging provided by PET or SPECT with anatomical imaging from computed tomography (CT) or magnetic resonance imaging (MRI) was soon recognized [56–58]. Anatomical imaging not only reveals the underlying morphology with great details but also allows the accurate localization of abnormal uptake in the functional PET or SPECT images, which is difficult otherwise due to low contrast and low resolution of anatomy provided by nuclear imaging.

The combined analysis of images from different modalities was first done visually. Later, software-based image registration was used, which provides satisfactory results for brain imaging but is challenging and error-prone for thoracic and abdominal regions. The limitations derived from the acquisition in different scanners and at different times triggered the investigation of hybrid scanners in the 1990s, first SPECT/CT [59, 60] and then PET/CT [61]. The first commercial hybrid scanner was the Hawkeye (GE Healthcare), introduced in 1999 [62]. The first Hawkeye was a SPECT system incorporating a low-power, non-diagnostic CT. Shortly after, in 2001, two combined PET/CT systems including a diagnostic CT were introduced: the Discovery LS (GE Healthcare) and the Biograph (CTI PET Systems, later part of Siemens Medical Solutions) [63], followed 1 year later by the Gemini (Philips Medical Systems). After a few years of discussion and divergent opinions, it can now be confidently claimed that hybrid systems such as PET/CT are able to provide significant clinical advantages [64, 65], while remaining cost-effective.

Since its introduction, hybrid systems have revolutionized nuclear imaging. The way to acquire, process and interpret the images has deeply changed, bringing important advances and meanwhile new challenges [66]. This impact has been particularly dramatic for PET: standalone PET in the 1990s was a modality rarely used clinically and dominated by research applications; since its integration into hybrid PET/CT, it has become a mainstream diagnostic imaging tool with a major role in patient management.

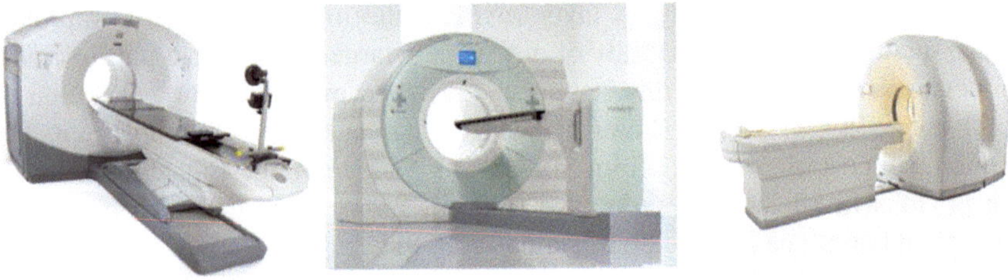

Fig. 4.7 Selected PET/CT designs: discovery (GE), Biograph (Siemens), and Gemini (Philips)

Moreover, hybrid scanners have also required a closer cooperation of two medical specialties: nuclear medicine and radiology. Health systems have had to adapt to this situation, finding solutions regarding the operation of the scanners and joint training and competence for the image reading and reporting of studies from hybrid systems [67, 68].

4.4.2 PET/CT

A hybrid PET/CT consists fundamentally of a standalone PET and a standalone CT scanner placed contiguous to each other, having a common patient bed, and with adapted acquisition and processing software (see Fig. 4.7). That is, the main architecture of both PET and CT is preserved in hybrid scanners, except for the disappearance of the rotating radionuclide source from PET because the attenuation correction is done based on the CT data.

Typically, a PET/CT examination starts with a topogram, an overview X-ray scan acquired in few seconds which is used to define the axial range of the subsequent CT and PET examinations. Then, the CT is acquired (helical acquisition with continuous bed motion in less than 20 s), followed by PET, using the step-and-shoot method in case of multi-bed acquisitions (with 1–4 min per bed position).

For the purpose of anatomical correlation and attenuation correction of the PET data, a low-dose unenhanced CT is sufficient. However, for some indications, a diagnostic CT enhanced with contrast agent might be done instead of or in addition to the low-dose CT (e.g., CT of the lungs during maximum inspiration for the detection of lung nodules or a CT angiography in cardiac exams) [69].

As mentioned, the CT data is used in PET/CT to determine the attenuation map to correct for photon attenuation. There are several advantages associated with the use of CT for attenuation correction instead of a rotating radionuclide source:

- Faster acquisition (shorter examination, improved throughput, and patient comfort)
- Lower statistical noise
- It can be acquired before or after administration of radiotracer without bias
- It does not need periodical replacement as the radionuclide rotating sources

CT-based attenuation correction is successfully used in clinical routine and has been validated even when low-tube currents are used for the CT acquisition [70, 71]. Yet, use of the CT for attenuation correction does not come without challenges. The attenuation map for PET should contain the linear attenuation factors for gamma rays at 511 keV, whereas the CT image shows the attenuation factors resulting from a polychromatic X-ray beam over a range of lower energies (40–140 keV). To convert from attenuation at CT energies to attenuation at PET energies, a bilinear scaling is used [72]. The attenuation maps generated using this approach are accurate in most cases, with two notable limitations:

- When metal objects are present in the FOV, since they create artifacts in the CT image which propagate into the PET data [73]
- When iodinated contrast agents are used in CT, since they have a high attenuation for X-rays but not for 511 keV photons [74]

A further challenge of using CT for attenuation correction of PET data is that CT is acquired within seconds and captures a fraction of the respiratory cycle, resulting in occasional misregistration between the PET and CT data which can lead to artifacts in the PET attenuation-corrected images [75, 76].

Because acquisition of the CT data is done immediately before or after acquisition of the PET data, a very good temporal and spatial correlation between the images is achieved. As described above, the functional information by PET and anatomical information by CT complement each other, and several studies have demonstrated an overall increased diagnostic accuracy of hybrid PET/CT when compared to analysis of separate PET and CT acquisitions [77].

In summary, PET/CT acquires faster than standalone PET, provides PET images with better quality (through attenuation maps with less noise), and adds the anatomical information provided by CT (Fig. 4.8). Clinical centers were soon convinced of the advantages of hybrid PET/CT: only 5 years after entering the market, PET/CT had made standalone PET scanners virtually obsolete.

4.4.3 SPECT/CT

In hybrid SPECT/CT, the CT image is also used beyond its own diagnostic value for anatomical referencing and attenuation correction of the SPECT data. The advantages of SPECT/CT over standalone SPECT are comparable to those for PET/CT versus standalone PET, except for the acquisition time, which is the same in SPECT/CT as in SPECT (because attenuation correction is not routinely performed in standalone SPECT). Yet, SPECT/CT has not had the same market penetration of PET/CT and still coexists with standalone SPECT scanners.

A reason for the less extended use of SPECT/CT as compared to PET/CT is its limited advantage—relative to its high cost—for SPECT's most frequent indication, myocardial perfusion imaging. In this context, SPECT/CT can add assessment of the coronary calcium (*calcium scoring*) and attenuation correction. The use of

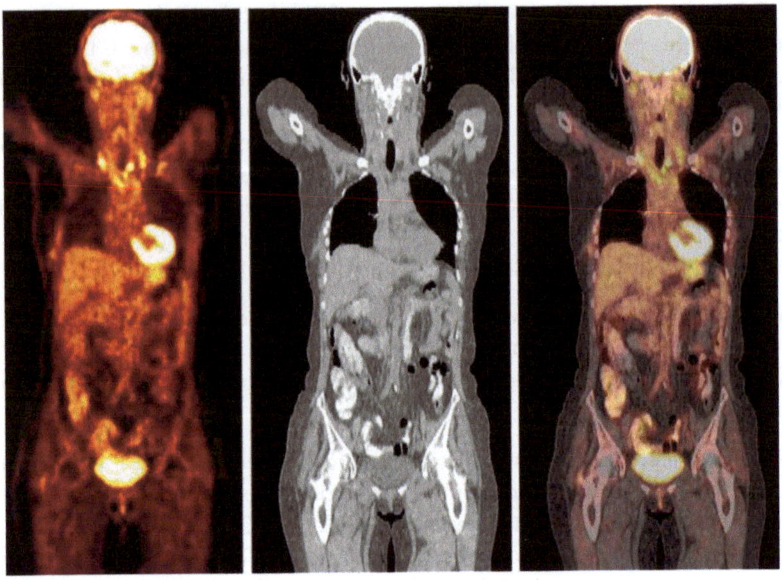

Fig. 4.8 Coronal views of ^{18}F-FDG PET, CT, and fused PET/CT from a patient with bronchial carcinoma examined in a hybrid PET/CT

attenuation for routine perfusion imaging with SPECT is still controversial [51]. Moreover, in order to provide the most valuable complementary information out of the CT data, a state-of-the-art CT scanner is necessary to do a CT angiography for evaluation of the coronary morphology; the cost-effectiveness of such configuration is questionable, because such a state-of-the-art CT will be underused when integrated into a SPECT/CT. Thus, when CT angiography is required, acquisition on a separate (state-of-the-art) CT is often preferred.

SPECT/CT provides a substantial advantage as compared to SPECT in other imaging tasks such as imaging of tumors, infections, and inflammations and optimized dosimetry for radiation treatment planning or radionuclide therapy [78–80]. The added clinical value of SPECT/CT because of attenuation correction and especially accurate anatomical co-localization in these applications has been well documented [81–84].

4.4.4 PET/MR

An initial motivation to combine PET and MRI was the possibility to use the strong magnetic field of MR to reduce the positron range in the tissues before annihilation. That would result in a significant improvement of the spatial resolution in PET [85]. However, it was soon determined that such gain in resolution would be insignificant in its expected clinical use for magnetic fields below 5 T and radioisotopes emitting low-energy positrons such as the most commonly used ^{18}F [86]. The gain in spatial resolution might be relevant mainly for high-resolution applications in high-field MRI (in small-animal imaging) and be more evident when using isotopes with high-energy positrons such as ^{82}Rb or ^{68}Ga.

In this context, small-animal PET/MR systems were investigated and developed [87, 88]. Following the success of PET/CT, which demonstrated the potential of hybrid imaging, there was an increasing interest in the development of clinical PET/MR systems. Nowadays, we are assisting at the birth of hybrid clinical PET/MR, with models offered already by two manufacturers (in both cases using a 3-T MRI). In hybrid PET/MR, MR can provide an anatomical reference for localization of PET positive lesions [89], just like CT in hybrid PET/CT. Beyond that, PET/MR has several potential advantages over PET/CT:

- *Improved soft tissue contrast.* MRI provides better tissue contrast than CT. Several tumor entities can be distinguished in MRI and not in CT.
- *Reduced exposure to ionizing radiation.* MRI uses no ionizing radiation, which is of particular importance for serial follow-up examinations and pediatric imaging [90].
- *True simultaneous acquisition.* As opposed to sequential acquisition in combined PET/CT, integrated PET/MR allows true simultaneous acquisition.
- *MR-based motion correction.* Due to the simultaneous acquisition and the lack of ionizing radiation, real-time MRI data can be used to account for the patient motion during the PET acquisition.
- *Advanced MRI sequences.* MRI offers a wide variety of measurements, including functional MRI, diffusion, and perfusion imaging as well as MR spectroscopy. These can add relevant information to the functional measurements by PET.
- *Decrease in positron range.* The positron range is reduced in the plane perpendicular to the main magnetic field, not in the direction parallel to it. As mentioned above, the resulting gain in in-plane spatial resolution will be clinically insignificant for low-energy isotopes such as ^{18}F but might be relevant for high-energy isotopes such as ^{82}Rb, especially if higher magnetic fields (e.g., 7 T) are used in the future.

Beside all these possible advantages, there is a major drawback for PET/MR systems. A conventional PET scanner cannot work within a strong magnetic field, because this field affects the functions of the photomultiplier tubes used for the light readout and render them inoperative. A solution proposed by the Ingenuity TF PET/MR (Philips) is to have sufficient distance between PET and MR machines and additional shielding to the PET components to avoid substantial interferences. This strategy preserves the conventional design and performance of the PET scanner (in particular its time resolution and time-of-flight capabilities), sacrificing the possibility to perform true simultaneous acquisitions (see Fig. 4.9). An alternative proposed by the Biograph mMR (Siemens) is to fully integrate the PET and MRI scanners concentrically in the same gantry using avalanche photodiodes as photodetectors instead of photomultiplier tubes. The operation of the photodiodes

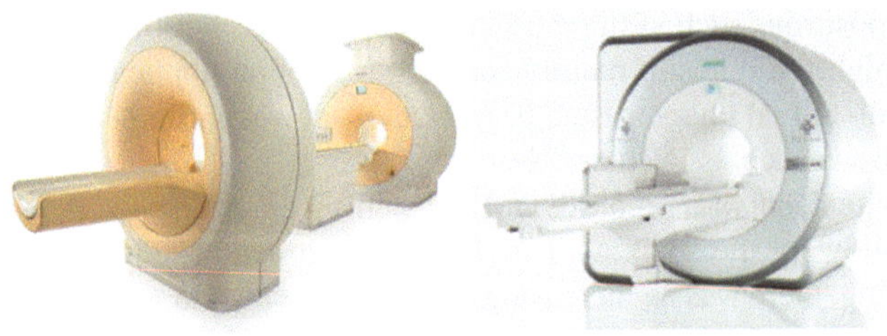

Fig. 4.9 First clinical PET/MR whole-body scanners proposing two different architectures: on the *left*, sequential design adopted in the Ingenuity TF PET/MR (Philips); on the *right*, integrated design, as adopted in the Biograph mMR (Siemens)

is not affected by the magnetic field, and their use allows true simultaneous acquisition and requires a smaller scanner room. However, a limited time resolution and the impossibility to make time-of-flight determinations have to be taken into account. A solution with the advantages of both approaches might come from the use of recently developed silicon photomultipliers for the light readout, which are MR-compatible and allow a time resolution sufficient for time-of-flight capabilities.

Another important challenge in PET/MR scanners is attenuation correction. As opposed to CT, MR does not provide direct information about the radiodensity of tissue. Therefore, a method to derive the attenuation factors out of MR images is needed. Initial investigation of this topic concentrated on brain imaging, registering the MR images to an atlas with well-known attenuation [91, 92]. For whole-body imaging, direct segmentation of the MR image in tissue classes has been proposed [93, 94]. MR-based attenuation correction faces the additional challenges of limited FOV in MRI [95] and attenuation produced by the MR coils [96].

Regarding its clinical use, hybrid PET/MR offers interesting possibilities in neurology [97], oncology [98], and cardiology [99]. Yet, protocols optimized for PET/MR still need to be developed, and it is still too early to draw conclusions about the diagnostic accuracy and cost-effectiveness of PET/MR versus PET/CT for different imaging tasks.

Acknowledgments The author thanks Isabel Dregely, Ph.D. and Shelley Zhang, Ph.D., for assistance in the final preparation of the manuscript.

References

1. Blumgart HL, Yens OC (1927) Studies on the velocity of blood flow: I. The method utilized. J Clin Invest 4:1–12
2. Loong CY, Anagnostopoulos C (2004) Diagnosis of coronary artery disease by radionuclide myocardial perfusion imaging. Heart 90(Suppl V):v2–v9
3. Kapur A, Latus KA, Davies G et al (2002) A comparison of three radionuclide myocardial perfusion tracers in clinical practice: the ROBUST study. Eur J Nucl Med Mol Imaging 29:1608–1616

4. Underwood SR, Anagnostopoulos C, Cerqueira M et al (2004) Myocardial perfusion scintigraphy: the evidence. Eur J Nucl Med Mol Imaging 31:261–291
5. Sciagra R, Pellegri M, Pupi A, Bolognese L, Bisi G, Carnovale V, Santoro GM (2000) Prognostic implications of Tc-99m sestamibi viability imaging and subsequent therapeutic strategy in patients with chronic coronary artery disease and left ventricular dysfunction. J Am Coll Cardiol 36:739–745
6. Camici PG, Prasad SK, Rimoldi OE (2008) Stunning, hibernation, and assessment of myocardial viability. Circulation 117:103–114
7. Holman BL, Tumeh SS (1990) Single-photon emission computed tomography (SPECT) applications and potential. JAMA 263:561–564
8. Meikle SR, Kench P, Kassiou M, Banati RB (2005) Small animal SPECT and its place in the matrix of molecular imaging technologies. Phys Med Biol 50:R45–R61
9. Franc BL, Acton PD, Mari C, Hasegawa BH (2008) Small-animal SPECT and SPECT/CT: important tools for preclinical investigation. J Nucl Med 49:1651–1663
10. Eisner RL, Nowak DJ, Pettigrew R, Fajman W (1986) Fundamentals of 180 degree acquisition and reconstruction in SPECT imaging. J Nucl Med 27:1717–1728
11. Moore SC, Kouris K, Cullum I (1992) Collimator design for single photon emission tomography. Eur J Nucl Med 19:138–150
12. Cullom SJ, Case JA, Bateman MD (1998) Electrocardiographically gated myocardial perfusion SPECT: technical principles and quality control considerations. J Nucl Cardiol 5:418–425
13. Paul AK, Nabi HA (2004) Gated myocardial perfusion SPECT: basic principles, technical aspects, and clinical applications. J Nucl Med Technol 32:179–187
14. Cho K, Kumiata SI, Okada S, Kumazaki T (1999) Development of respiratory gated myocardial SPECT system. J Nucl Cardiol 6:20–28
15. Gullberg GT, Reutter BW, Sitek A, Maltz JS, Budinger TF (2010) Dynamic single photon emission computed tomography—basic principles and cardiac applications. Phys Med Biol 55: R111–R191
16. Cao Z, Maunoury C, Chen CC, Holder LE (1996) Comparison of continuous step-and-shoot versus step-and-shoot acquisition SPECT. J Nucl Med 37:2037–2040
17. Gottschalk SC, Salem D, Lim CB, Wake RH (1983) SPECT resolution and uniformity improvements by noncircular orbit. J Nucl Med 24:822–828
18. Todd-Pokropek A (1983) Non-circular orbits for the reduction of uniformity artefacts in SPECT. Phys Med Biol 28:309–313
19. Beekman F, van der Have F (2007) The pinhole: gateway to ultra-high-resolution three-dimensional radionuclide imaging. Eur J Nucl Med Mol Imaging 34:151–161
20. Funk T, Després P, Barber WC, Shah KS, Hasegawa BH (2006) A multipinhole small animal SPECT system with submillimeter spatial resolution. Med Phys 33:1259–1268
21. van der Have F, Vastenhouw B, Ramakers RM et al (2009) U-SPECT-II: an ultra-high-resolution device for molecular small-animal imaging. J Nucl Med 50:599–605
22. Bateman TM, Cullom SJ (2005) Attenuation correction single-photon emission computed tomography myocardial perfusion imaging. Semin Nucl Med 35:37–51
23. Bruyant PP (2002) Analytic and iterative reconstruction algorithms in SPECT. J Nucl Med 43:1343–1358
24. Sharir T, Slomkam PJ, Bermanm DS (2010) Solid-state SPECT technology: fast and furious. J Nucl Cardiol 17:890–896
25. Gambhir SS, Berman DS, Ziffer J (2009) A novel high-sensitivity rapid-acquisition single-photon cardiac imaging camera. J Nucl Med 50:635–643
26. Erlandsson K, Kacpersk K, van Gramberg D, Hutton BF (2009) Performance evaluation of D-SPECT: a novel SPECT system for nuclear cardiology. Phys Med Biol 54:2635–2649
27. Bocher M, Blevis IM, Tsukerman L, Shrem Y, Kovalski G, Volokh L (2010) A fast cardiac gamma camera with dynamic SPECT capabilities: design, system validation and future potential. Eur J Nucl Med Mol Imaging 37:1887–1902

28. Strauss LG (1997) Positron emission tomography: current role for diagnosis and therapy monitoring in oncology. Oncologist 2:381–388
29. Bomanji JB, Costa DC, Ell PJ (2001) The clinical role of positron emission tomography. Lancet Oncol 3:157–64
30. Basu S, Alavi A (2008) Unparalleled contribution of ^{18}F-FDG PET to medicine over 3 decades. J Nucl Med 49:17N–21N
31. Fletcher JW, Djulbegovic B, Soares HP et al (2008) Recommendations on the use of ^{18}F-FDG PET in oncology. J Nucl Med 49:480–508
32. Delbeke D, Coleman RE, Guiberteau MJ et al (2006) Procedure guideline for tumor imaging with ^{18}F-FDG PET/CT 1.0. J Nucl Med 47:885–895
33. Gillies RJ, Robey I, Gatenby RA (2008) Causes and consequences of increased glucose metabolism of cancers. J Nucl Med 49:24S–42S
34. Kumar R, Dhanpathi H, Basu S, Rubello D, Fanti S, Alavi A (2008) Oncologic PET tracers beyond [^{18}F]FDG and the novel quantitative approaches in PET imaging. Q J Nucl Med Mol Imaging 52:50–65
35. Bengel FM, Higuchi T, Javadi MS, Lautamäki R (2009) Cardiac positron emission tomography. J Am Coll Cardiol 54:1–15
36. Tai YF, Piccini P (2004) Applications of positron emission tomography (PET) in neurology. J Neurol Neurosurg Psychiatry 75:669–676
37. Nanni C, Rubello D, Fanti S (2007) Role of small animal PET for molecular imaging in preclinical studies. Eur J Nucl Med Mol Imaging 34:1819–1822
38. Casey ME, Nutt RA (1986) A multicrystal, two-dimensional BGO detector system for positron emission tomography. IEEE Trans Nucl Sci 33:460–463
39. Badawi RD, Marsden PK, Cronin BF, Sutcliffe JL, Maisey MN (1996) Optimization of noise-equivalent count rates in 3D PET. Phys Med Biol 41:1755–1776
40. Bailey DL (2003) Data acquisition and performance characterization in PET. In: Valk PE, Bailey DL, Townsend DW, Maisey MN (eds) Positron emission tomography: basic science and clinical practice. Springer, London, pp 41–62
41. Dahlbom M, Yu DC, Cherry SR, Chatziioannou A, Hoffman EJ (1992) Methods for improving image quality in whole body PET scanning. IEEE Trans Nucl Sci 39:1079–1083
42. Dahlbom M, Reed J, Young J (2001) Implementation of true continuous bed motion in 2-D and 3-D whole-body PET scanning. IEEE Trans Nucl Sci 48:1465–1469
43. Meikle SR, Badawi RD (2003) Quantitative techniques in PET. In: Valk PE, Bailey DL, Townsend DW, Maisey MN (eds) Positron emission tomography: basic science and clinical practice. Springer, London, pp 41–62
44. Nehmeh SA, Erdi YE, Ling CC et al (2002) Effect of respiratory gating on quantifying PET images of lung cancer. J Nucl Med 43:876–881
45. Martinez-Möller A, Zikic D, Botnar R et al (2007) Dual cardiac-respiratory gated PET: implementation and results from a feasibility study. Eur J Nucl Med Mol Imaging 34:1447–1454
46. Kaufmann PA, Camici PG (2005) Myocardial blood flow measurement by PET: technical aspects and clinical applications. J Nucl Med 46:75–88
47. Conti M (2011) Focus on time-of-flight PET: the benefits of improved time resolution. Eur J Nucl Med Mol Imaging 38:1147–1157
48. Lecomte R (2009) Novel detector technology for clinical PET. Eur J Nucl Med Mol Imaging 36:69–85
49. Degenhardt C, Prescher G, Frach T, Thon A, de Gruyter R, Schmitz A, Ballizany R (2009) The digital silicon photomultiplier—a novel sensor for the detection of scintillation light. IEEE Nucl Sci Symp Conf Rec 2383–2386
50. Ben-Haim S, Kacperski K, Hain S et al (2010) Simultaneous dual-radionuclide myocardial perfusion imaging with a solid-state dedicated cardiac camera. Eur J Nucl Med Mol Imaging 37:1710–1721

51. Germano G, Slomka PJ, Berman DS (2007) Attenuation correction in cardiac SPECT: the boy who cried wolf? J Nucl Cardiol 14:25–35
52. Alavi A, Basu S (2008) Planar and SPECT imaging in the era of PET and PET–CT: can it survive the test of time? Eur J Nucl Med Mol Imaging 35:1554–1559
53. Jansen FP, Vanderheyden JL (2007) The future of SPECT in a time of PET. Nucl Med Biol 34:733–735
54. Gholamrezanehad A, Mirpour S, Mariani G (2009) Future of nuclear medicine: SPECT versus PET. J Nucl Med 50:16N–18N
55. Mariani G, Strauss HW (2011) Positron emission and single-photon emission imaging: synergy rather than competition. Eur J Nucl Med Mol Imaging 38:1189–90
56. Schad LR, Boesecke R, Schiegel W et al (1987) Three-dimensional image correlation of CT, MR and PET studies in radiotherapy treatment planning of brain tumors. J Comput Assist Tomogr 11:948–954
57. Pelizzari CA, Chen GTY, Spelbring DR, Weichselbaum RR, Chen CT (1989) Accurate three-dimensional registration of CT, PET and/or MR images of the brain. J Comput Assist Tomogr 13:20–26
58. Ito K, Kato T, Tadokoro M, Ishiguchi T, Oshima M, Ishigaki T, Sakuma S (1992) Recurrent rectal cancer and scar: differentiation with PET and MR imaging. Radiology 182:549–552
59. Hasegawa BH, Stebler B, Rutt BK et al (1991) A prototype high-purity germanium detector system with fast photon-counting circuitry for medical imaging. Med Phys 18:900–909
60. Lang TF, Hasegawa BH, Liew SC, Brown JK, Blankespoor SC, Reilly SM, Gingold EL, Cann CE (1992) Description of a prototype emission transmission computed tomography imaging system. J Nucl Med 33:1881–1887
61. Beyer T, Townsend DW, Brun T et al (2000) A combined PET/CT scanner for clinical oncology. J Nucl Med 41:1369–1379
62. Patton JA, Delbeke D, Sandler MP (2000) Image fusion using an integrated, dual-head coincidence camera with X-ray tube-based attenuation maps. J Nucl Med 41:1364–1368
63. Townsend DW, Beyer T (2002) A combined PET/CT scanner: the path to true image fusion. Br J Radiol 75:24–30
64. Even-Sapir E, Keidar Z, Bar-Shalom R (2009) Hybrid imaging (SPECT/CT and PET/CT)—improving the diagnostic accuracy of functional/metabolic and anatomic imaging. Semin Nucl Med 39:264–275
65. Delbeke D, Schöder H, Martin WH, Wahl RL (2009) Hybrid imaging (SPECT/CT and PET/CT): improving therapeutic decisions. Semin Nucl Med 39:308–340
66. Martinez-Möller A (2009) Challenges in multimodality imaging using positron emission tomography. Ph.D. Dissertation, Technische Universität München, Germany
67. Fischman AJ, Thrall JH (2003) Who should read and interpret ^{18}F-FDG PET studies? J Nucl Med 44:1197–1199
68. Bischof Delaloye A, Carrió I, Cuocolo A et al (2007) White paper of the European Association of Nuclear Medicine (EANM) and the European Society of Radiology (ESR) on multimodality imaging. Eur J Nucl Med Mol Imaging 34:1147–1151
69. Antoch G, Freudenberg LS, Beyer T, Bockisch A, Debatin JF (2004) To enhance or not to enhance? ^{18}F-FDG and CT contrast agents in dual-modality ^{18}F-FDG PET/CT. J Nucl Med 45 (Suppl):56–65
70. Burger C, Goerres G, Schoenes S, Buck A, Lonn AH, von Schulthess GK (2002) PET attenuation coefficients from CT images: experimental evaluation of the transformation of CT into PET 511-keV attenuation coefficients. Eur J Nucl Med Mol Imaging 29:922–927
71. Souvatzoglou M, Bengel F, Busch R, Kruschke C, Fernolendt H, Lee D, Schwaiger M, Nekolla SG (2007) Attenuation correction in cardiac PET/CT with three different CT protocols: a comparison with conventional PET. Eur J Nucl Med Mol Imaging 34:1991–2000
72. Kinahan PE, Townsend DW, Beyer T, Sashin D (1998) Attenuation correction for a combined 3D PET/CT scanner. Med Phys 25:2046–2053

73. Goerres GW, Ziegler SI, Burger C, Berthold T, von Schulthess GK, Buck A (2003) Artifacts at PET and PET/CT caused by metallic hip prosthetic material. Radiology 226:577–584
74. Mawlawi O, Erasmus JJ, Munden RF et al (2006) Quantifying the effect of IV contrast media on integrated PET/CT: clinical evaluation. Am J Roentgenol 186:308–319
75. Beyer T, Antoch G, Blodgett T, Freudenberg LF, Akhurst T, Mueller S (2003) Dual-modality PET/CT imaging: the effect of respiratory motion on combined image quality in clinical oncology. Eur J Nucl Med Mol Imaging 30:588–596
76. Martinez-Möller A, Souvatzoglou M, Navab N, Schwaiger M, Nekolla SG (2007) Artifacts from misaligned CT in cardiac perfusion PET/CT studies: frequency, effects, and potential solutions. J Nucl Med 48:188–193
77. Czernin J, Allen-Auerbach M, Schelbert H (2007) Improvements in cancer staging with PET/CT: literature-based evidence as of September 2006. J Nucl Med 48(Suppl 1):78–88
78. Römer W, Nomayr A, Uder M, Bautz W, Kuwert T (2006) SPECT-guided CT for evaluating foci of increased bone metabolism classified as indeterminate on SPECT in cancer patients. J Nucl Med 47:1102–1106
79. Buck AK, Nekolla SG, Ziegler SI et al (2008) SPECT/CT. J Nucl Med 49:1305–1319
80. Mariani G, Bruselli L, Kuwert T, Kim EE, Flotats A, Israel O, Dondi M, Watanabe N (2010) A review on the clinical uses of SPECT/CT. Eur J Nucl Med Mol Imaging 37:1959–1985
81. Utsunomiya D, Shiraishi S, Imuta M et al (2006) Added value of SPECT/CT fusion in assessing suspected bone metastasis: comparison with scintigraphy alone and nonfused scintigraphy and CT. Radiology 238:264–271
82. Tharp K, Israel O, Hausmann J et al (2004) Impact of ^{131}I-SPECT/CT images obtained with an integrated system in the follow-up of patients with thyroid carcinoma. Eur J Nucl Med Mol Imaging 31:1435–1442
83. Lavely WC, Goetze S, Friedman KP et al (2007) Comparison of SPECT/CT, SPECT, and planar imaging with single- and dual-phase ^{99m}Tc-sestamibi parathyroid scintigraphy. J Nucl Med 48:1084–1089
84. Gayed IW, Kim EE, Broussard WF et al (2005) The value of ^{99m}Tc-sestamibi SPECT/CT over conventional SPECT in the evaluation of parathyroid adenomas or hyperplasia. J Nucl Med 46:248–252
85. Hammer BE, Christensen NL, Heil BG (1994) Use of a magnetic field to increase the spatial resolution of positron emission tomography. Med Phys 21:1917–1920
86. Raylman RR, Hammer BE, Christensen NL (1996) Combined MRI-PET scanner: a Monte Carlo evaluation of the improvements in PET resolution due to the effects of a static homogeneous magnetic field. IEEE Trans Nucl Sci 43:2406–2412
87. Shao Y, Cherry SR, Farahani K et al (1997) Development of a PET detector system compatible with MRI/NMR systems. IEEE Trans Nucl Sci 44:1167–1171
88. Judenhofer MS, Wehrl HF, Newport DF et al (2008) Simultaneous PET-MRI: a new approach for functional and morphological imaging. Nat Med 14:459–465
89. Eiber M, Martinez-Möller A, Souvatzoglou M et al (2011) Value of a Dixon based MR-PET attenuation correction sequence for the localization and evaluation of PET positive lesions. Eur J Nucl Med Mol Imaging 38:1691–1701
90. Brix G, Nekolla EA, Nosske D, Griebel J (2009) Risks and safety aspects related to PET/MR examinations. Eur J Nucl Med Mol Imaging 36(Suppl 1):131–138
91. Rota Kops E, Herzog HR (2008) Template-based attenuation correction of PET in hybrid MR-PET [abstract]. J Nucl Med 49(Suppl):162P
92. Hofmann M, Steinke F, Scheel V et al (2008) MRI-based attenuation correction for PET/MRI: a novel approach combining pattern recognition and atlas registration. J Nucl Med 49:1875–1883
93. Martinez-Möller A, Souvatzoglou M, Delso G et al (2009) Tissue classification as a potential approach for attenuation correction in whole-body PET/MRI: evaluation with PET/CT data. J Nucl Med 50:520–526

94. Schulz V, Torres-Espallardo I, Renisch S et al (2011) Automatic, three-segment, MR-based attenuation correction for whole-body PET/MR data. Eur J Nucl Med Mol Imaging 38:138–152
95. Delso G, Martinez-Möller A, Bundschuh R, Nekolla SG, Ziegler SI (2010) The effect of limited MR field of view in MR/PET attenuation correction. Med Phys 37:2804–2812
96. Delso G, Martinez-Möller A, Bundschuh R, Ladebeck R, Candidus Y, Faul D, Ziegler SI (2010) Evaluation of the attenuation properties of MR equipment for its use in a whole-body PET/MR scanner. Phys Med Biol 55:4361–4374
97. Heiss W-D (2009) The potential of PET/MR for brain imaging. Eur J Nucl Med Mol Imaging 36(Suppl 1):105–112
98. Antoch G, Bockisch A (2009) Combined PET/MRI: a new dimension in whole-body oncology imaging? Eur J Nucl Med Mol Imaging 36(Suppl 1):113–120
99. Nekolla SG, Martinez-Möller A, Saraste A (2009) PET and MRI in cardiac imaging: from validation studies to integrated applications. Eur J Nucl Med Mol Imaging 36(Suppl 1):121–130

Chapter 5
Perspectives in Nuclear Medicine Tomography: A Physicist's Point of View

Stephan Walrand and François Jamar

5.1 Ultimate Goal: Goodbye Prof. Radon, Welcome Back Prof. Poisson

A major contribution of physicists in nuclear medicine, jointly with mathematicians, was to invent and develop dedicated tomographic reconstruction methods. However, the ultimate dream of each physicist is to develop acquisition systems recording sufficient information for each detected event to allow the direct determination of the decay location that generated this event. Two different ways are possible to achieve this ultimate goal: (1) in TOF-PET, by an accurate measure of the delay separating the detection of the two coincident collinear 511 keV γ-rays, and (2) in SPECT, by the measure of the incoming direction of two coincident noncollinear γ-rays. Obviously, this last method requires isotopes owning a γ-ray cascade in their decay scheme.

Even if the determination of the decay location event per event requires some computational treatment of the acquired information, i.e., the time-of-flight or the incoming directions of the detected γ-rays, this treatment is definitely not a reconstruction process. Reconstruction in conventional tomography consists in determining for all measured counts the fraction of γ-rays that has to be spread on each voxel owned by the line of response (LOR). Individually, these measured counts do not provide any information about this spread along the LOR. The only way is to determine the spread that is compatible with all the measured LORs. In 1917, Johan Radon proved such determination feasible using an analytical algorithm. A major drawback of conventional tomography is that several different spreads are compatible with the measured data, resulting in what is called an ill-posed inverse problem. As a result, despite the development of very sophisticated iterative algorithms to

S. Walrand (✉) • F. Jamar
Nuclear medicine center, Université Catholique de Louvain, Brussels, Belgium
e-mail: Stephan.walrand@uclouvain.be

A. Giussani and C. Hoeschen (eds.), *Imaging in Nuclear Medicine*,
DOI 10.1007/978-3-642-31415-5_5, © Springer-Verlag Berlin Heidelberg 2013

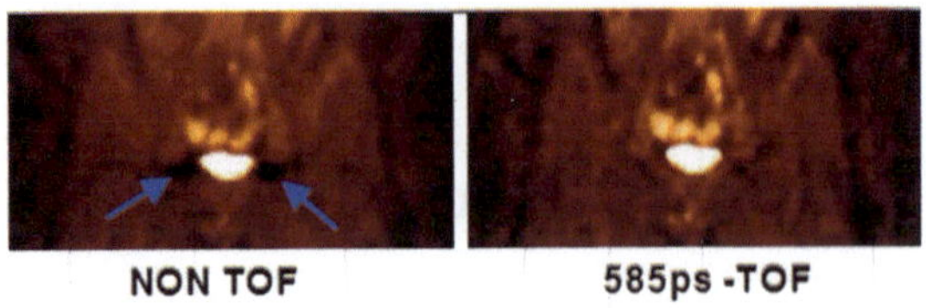

Fig. 5.1 Reduction of the artifacts present around the bladder in non-TOF reconstruction. Reprinted from [1] with permission of the Society of Nuclear Medicine

regularize the problem, noise amplification and artifact generation during the reconstruction process are still a major issue (Fig. 5.1).

Current commercial TOF-PET systems partially address this issue: the TOF information reduces the noise propagation during the reconstruction process by localizing each event in a LOR segment (prelocalization) rather than spreading statistical noise along the full length of the LOR [2]. Campbell et al. [3] quantified the improvement provided by the TOF information in estimating the noise reduction factor f obtained for the reconstruction process as a function of the time resolution (TR) and of the diameter of the body imaged (Table 5.1).

There is another aspect of the improvement rarely discussed that concerns an interesting benefit to improve the decay prelocalization: the modification of the noise properties in the reconstructed image. In planar acquisition the noise in the image follows a Poisson distribution and is spatially uncorrelated, i.e., the noise affecting a pixel depends only on the value of this pixel and its variance is equal to the square root of this value. In tomography, the activity in a reconstructed voxel is obtained from the measurement of the total activity on all the LORs traversing that voxel. Consequently, the noise does not longer follow a Poisson distribution and is furthermore spatially correlated, i.e., the noise affecting a voxel also depends on the values of the other voxels in the image. This explains why the planar images usually appeared more natural and more comfortable for the human vision. Lastly, this spatial correlation induces artifacts, especially in the neighborhood of highly active organs (Fig. 5.1). The prelocalization provided by TOF reduces this spatial correlation and the artifact generation. For an ultimate time resolution lower than 25 ps, TOF-PET will directly localize the annihilation in the correct 4-mm voxel without any reconstruction process. At that time, the noise in the tomographic image will follow the Poisson law and will be spatially uncorrelated, such as in planar image.

5.2 New Technology Developments in SPECT

5.2.1 More and More Solid: End of the PMT Age

The photomultiplier tube (PMT) was invented in 1934 by Harley Iams and Bernard Salzberg who joined together in the same vacuum housing the photoelectric effect discovered in 1887 by Heinrich Hertz (explained in 1905 by Albert Einstein) and the amplification tube invented in 1919 by Joseph Slepian (Fig. 5.2). The birth and

Table 5.1 Noise reduction factor f obtained in function of the time resolution (TR) of the TOF-PET and of the diameter of the body (20 and 35 cm)

		f	
TR (ps)	PL (mm)	20 cm	35 cm
1,000	150	1.3	2.3
500	75	2.7	4.7
20	3	66.7	117.0

PL is the FWHM of the prelocalization provided by the TOF information
Reprinted with permission from [3] courtesy of Dr. Robert Kowalewski and Dr. Warren Campbell (University of Victoria, CA)

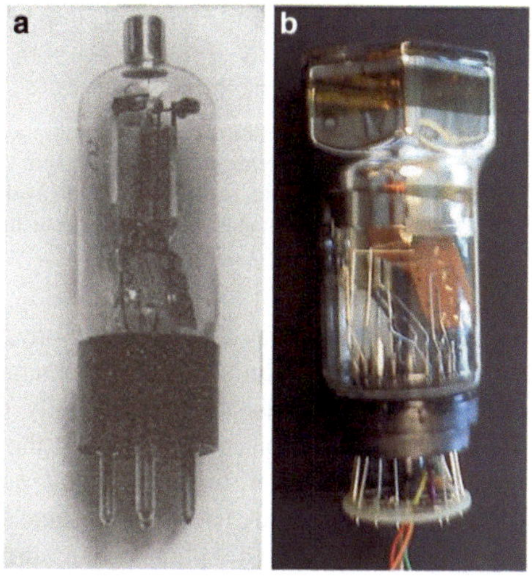

Fig. 5.2 (**a**) First PMT invented by Iams and Salzberg (reprinted from [4] with permission of IEEE). (**b**) Modern buttable PMT used in gamma camera

the development of the nuclear medicine are so much linked to the PMT to pay tribute to these great scientists.

In 1947 the transistor, a solid-state version of the amplification tube, was invented by John Bardeen, William Shockley, and Walter Brattain at the Bell Company. Solid-state components have structural benefits versus vacuum glass tubes: they are insensitive to magnetic fields, they are more robust, they have a longer life, their performances do not change with age, they are more compact, and they don't require a high voltage source. Although the transistor performances were still below those of the amplification tube, these structural benefits were already sufficient for Sony to introduce the first whole transistor radio in 1955.

The commercial success of an electronic component requires two distinct steps: (1) the invention and performance improvement of the component and (2) the development of an industrial process to reproducibly manufacture the component at an attractive cost. This second step is often the most expensive step. All electronic devices require signal amplification, and the amplification tube got a mass success with the worldwide spreading of the television in the beginning of the golden 1960s. This growing large market motivated major electronic device

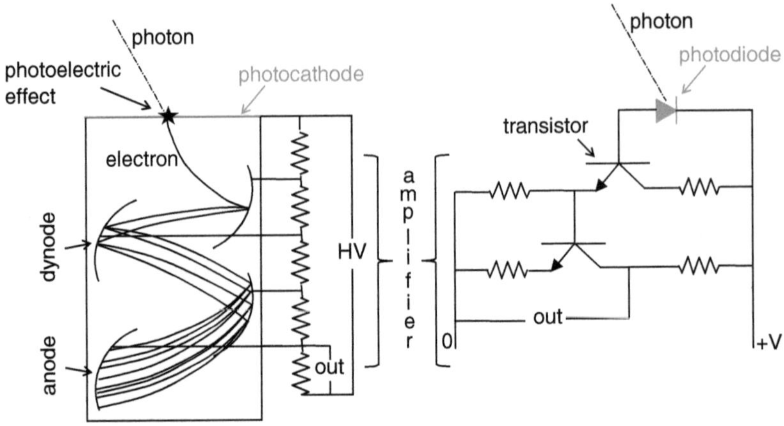

Fig. 5.3 Schematic diagram of the photomultiplier tube (*left*) and of its solid-state replacement (*right*). Photo sensitive areas are about $\pi\ 30^2\ \mathrm{mm}^2$ and about $2 \times 2\ \mathrm{mm}^2$ for the photomultiplier tube and for the photodiode, respectively. With avalanche photodiode, the major part of the gain is already provided by the avalanche process occurring inside the photodiode

manufacturers, such as Bell, RCA, Sony, and Philips, to focus their efforts in improving the noise and speed performance of transistors. During the 1970s the performances of transistors dramatically overpassed those of the amplification tubes. Transistors replaced tubes in all the existing applications and opened new fields such as computer, high-frequency satellite communications.

In contrast, PMTs are used only in some scientific applications and rarely in a tile assembly. Although the photovoltaic property of the p-n junction had been observed in 1940 by Russell Ohl already 7 years before it was used in the transistor, comparatively less efforts were spent by the major manufacturers to develop a reproducible and commercially production process of four-side buttable solid-state components to replace the PMT. Nowadays 98 % of the nuclear medicine system is still based on the PMT scintillation counter invented in 1944 by Samuel Curran, and the development of solid-state replacement components is still mainly the job of "small" companies.

Two solid-state components emerged to replace the PMT: the avalanche photodiode (APD) (Fig. 5.3) and the Geiger-mode avalanche photodiode (G-APD) that can detect single photon just like a PMT with a very high gain and therefore is often called silicon photomultiplier (SiPM). At the same time, besides the pure replacement of the PMT, the cadmium zinc telluride (CZT) detector that replaces the crystal-PMT tandem was also developed. CZT can be seen as a kind of photodiode directly sensitive to γ-rays.

As shown in Table 5.2, solid-state detectors are still expensive, and their performances begin to reach, but not to clearly overpass, those of the PMTs. The wide introduction of the solid-state components in nuclear medicine imaging will proceed in three phases similarly to the transistor story.

Table 5.2 Photodetector properties extracted from [5]

	PMT	APD	SiPM	CZT
Gain	**10^7**	10^3	**10^6**	**n.a.**
Noise	**Low**	Moderate	**Low**	**Low**
Threshold (ph.e)	**1**	10	**1**	**n.a.**
Timing resolution	**0.5 ns**	1 ns	**0.5 ns**	7 ns
Energy resolution	10 %	13 %	10 %	**4 %**
Cost	**$**	$$	$$$$	$$$
MR compatible	No	**Yes**	**Yes**	**Yes**

Strengths of the components are highlighted in bold

The first phase, equivalent to the introduction of the whole transistor radio in 1955, already started a few years ago. Although expensive, the structural properties of the solid-state detectors offer potential for new acquisition techniques. The compactness significantly reduces the size and the weight of the detector by decreasing the area of the shielding housing, thus resulting in a simplification of the gantry mechanic. In addition, contrary to the Anger camera that has a large dead area on the four edges of the crystal, all the CZT array surface is usable. The dead area of the detector is thus reduced to the housing thickness, i.e., less than 1 cm using tungsten as shielding material. These features allow static small CZT detectors to be set close together and focusing to a common target, as in the GE cardiac UFC [6]. Faster and more sophisticated motions can also be introduced, such as those used in the recently commercialized cardiac CZT D-SPECT camera (Spectrum Dynamics, Haifa, Israel) (Fig. 5.4). Lastly, the insensitivity to magnetic fields opens the development of simultaneous PET-MRI acquisition systems (as mentioned in Chap. 4).

The second phase, equivalent to the use of the transistor in the television, will start when the CZT cost will approach the crystal-PMT tandem cost. At this time, the better energy resolution provided by the CZT will on its own justify its use in all clinical SPECT systems: by reducing the scatter contribution, it will improve the image contrast, especially in heavy patient for whom the SPECT quality is the lowest.

The last phase, equivalent to the development of any new technology such as the computer, will occur when the performances of the solid-state detectors will overpass those of the PMTs and become sufficient for new acquisition concepts such as efficient Compton camera or positron emission locator. These concepts will be discussed in the following sections.

5.2.2 Lighter and Faster: The Compton Camera

Electronic collimation is the feature that, by avoiding the mechanical collimator, makes PET sensitivity and spatial resolution better than SPECT by a factor of about 100 and 3, respectively. Differently from the annihilation γ-rays produced by positron emitters, the direction of the multiples γ-rays emitted by the other isotopes

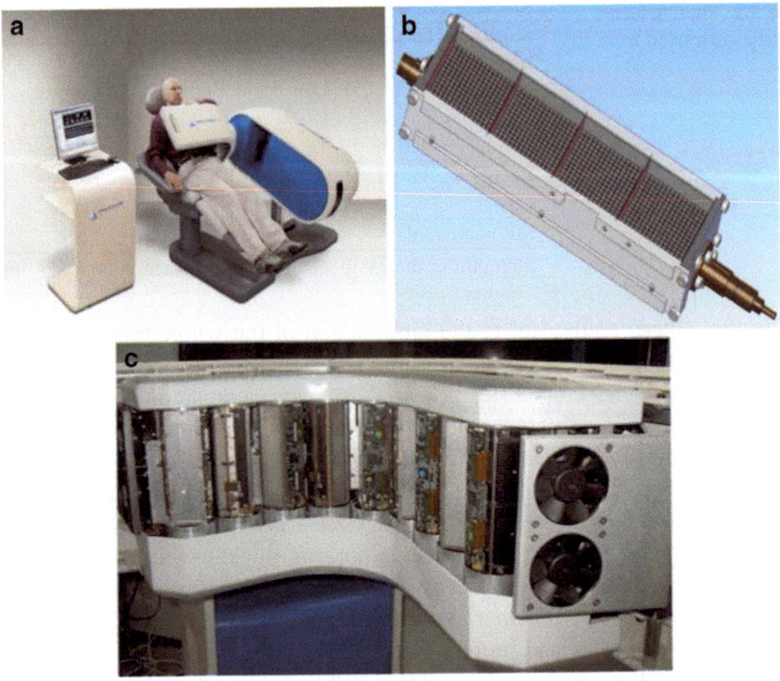

Fig. 5.4 Cardiac D-SPECT camera made of nine fast rotating compact column detectors (**c**) made of CZT detectors (**b**) scanning the heart region (**a**). (**b**) shows a column detector without its collimator; all the CZT pixels up to the housing edge are used. Reprinted from [6] with permission of Springer

is not strongly correlated. Thus, when a γ-ray is recorded by a collimated detector (typical efficiency $\approx 10^{-4}$), the probability that another γ-ray, originating from the same decay, has the right direction to be recorded by this detector, or another collimated detector, is almost zero (e.g., $\approx 10^{-8}$ to detect the two γ-rays emitted by ^{111}In). As a result, only one γ-ray is recorded per decay, justifying the name "single-photon emission" for the method. An alternative way to obtain information about the incoming direction of a single photon, besides the use of a mechanical collimator to remove the photons of unwanted direction, is to measure the crossing point of the photons trajectory in at least two different planes.

Already in 1974 Todd et al. [7] proposed to use two detectors in series to perform an electronic collimation in a single-photon camera and to perform 3D acquisition such as in PET. The first detector is optimized to favor Compton scattering, while the second is a classical γ camera that measures both the crossing point of the trajectory and the residual energy of the recoil photon. Knowing the energy of the primary photon and measuring the energy of the scattered photon, it is possible to determine the scattering angle θ and therefore the cone surface where the source is located. However, after one-third of century of research, no compact functional Compton camera has yet been produced due to the lack of a suitable Compton

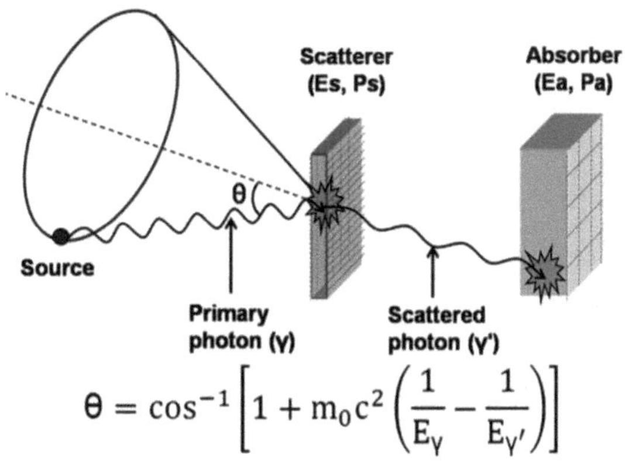

Fig. 5.5 Schematic rendering and principle of a Compton gamma camera. Reprinted from [9] with permission of Elsevier Ltd

detector. Such a detector has to be as large as possible and of high density in order to provide good efficiency and a high spatial resolution, it has to be made with material with low Z to favor the Compton scattering, and the electronic layers behind the detection volume area, if present, have to be fully transparent to the scattered photon to avoid additional Compton scatterings and thus supplementary changes in direction. This last condition has prevented the use of PMTs that up to recent years was the only affordable way to build large area efficient detectors.

Even using APD or CZT, the construction of such large Compton detector with hundreds of thousands detection elements, each with a volume less than $2 \times 2 \times 2$ mm^3, is difficult since each element requires high-resolution readout channels. The recent development of new application-specific integrated circuits (ASIC) partially addresses this issue [8]. Another way is to use double-sided skew-strip segmented planar silicon [9–12] or nitrogen-cooled germanium [13] detectors. Using these devices, the readout electronics is located on the edge of the detector area, outside the photon trajectories.

The Compton relation describing the photon energy conservation joined with the measure of the two crossing points of the trajectory in the two detector planes and of the residual energy of the recoil photon gives the equation of the cone in which the decay has occurred (Fig. 5.5). It is worth noting that even if a high improvement in sensitivity could be obtained, the situation is however not so favorable as that in PET, where the electronic collimation determines a single LOR or, more exactly, a thin cylinder due to limitation in spatial resolution. In Compton camera, the electronic collimation gives a cone surface that can have a large aperture angle θ (Fig. 5.5) and also with some thickness due to limitation in energy resolution.

5.2.3 *The Recoil Electron-Tracking Compton Camera*

Kabuki et al. [14, 15] proposed an elegant method to achieve the determination of a single LOR for a Compton camera, such as in PET. Beside the energy conservation,

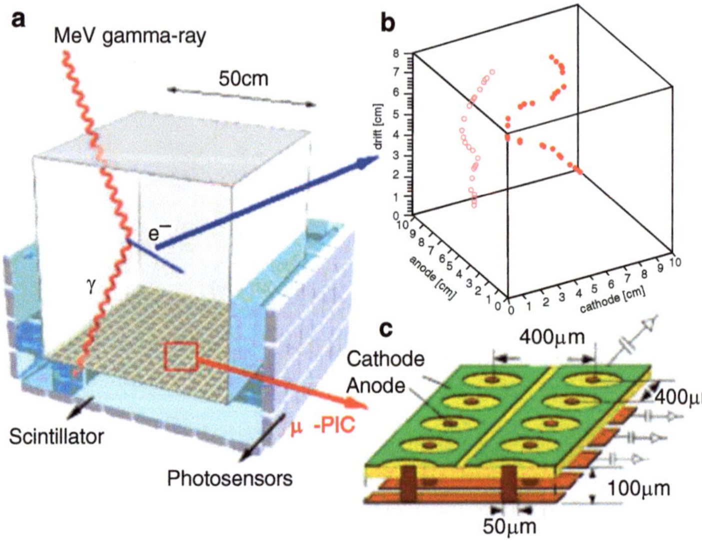

Fig. 5.6 Schematic rendering and principle of a recoil electron-tracking Compton gamma camera. (**a**) Gas detector. The scattered gamma-ray is detected using the scintillator-photosensor tandem, and the recoil electron is tracked by the μ-PIC detector. (**b**) Typical electron paths. (**c**) Expanded view of the μ-PIC detector. Reprinted from [14] with permission of Elsevier Ltd

momentum is also conserved in elastic collisions. Linear momentum is a vector defined by the direction of the particle motion. Its module for matter is the particle mass times the velocity and, for a photon, as observed by Arthur Compton in 1922, is the energy divided by the speed of light, i.e., E/c. Due to the law of conservation of linear momentum, the trajectories of the incident photon, of the recoil photon, and of the recoil electron are coplanar. Using a micropixel high-pressure gas chamber, Kabuki et al. built a Compton camera that enabled to track also the recoil electron path (Fig. 5.6).

With this device, the location of the decay is then restricted to the two lines defined by the intersection of the cone (deduced from the recoil photon analysis) with the plane defined by the trajectories of the scattered photon and of the recoil electron. Only one of these two lines satisfies the vectorial law of the momentum conservation, resulting in a single LOR. A first image of a thyroid phantom was obtained using this technique (Fig. 5.7, right). The low quality of the image does not prejudice the final quality that it will ultimately be possible to reach. For example, Fig. 5.7 (left) shows the first image of a thyroid obtained using a PMT-NaI detector that has been the state-of-the-art detection technique during 50 years.

Contrary to the Anger camera, a Compton camera has a wide energy dynamic, ranging from 100 to 3,000 keV. This opens the use of high-energy emitters such as ^{83}Sr (33h), ^{198}Tl(7h), ^{199}Tl(7h), ^{204}Bi(11h), ^{205}Bi(360h), ^{95}Tc (20h), and ^{96}Tc(103h). The two technetium isotopes could be particularly interesting for assessing slow processes.

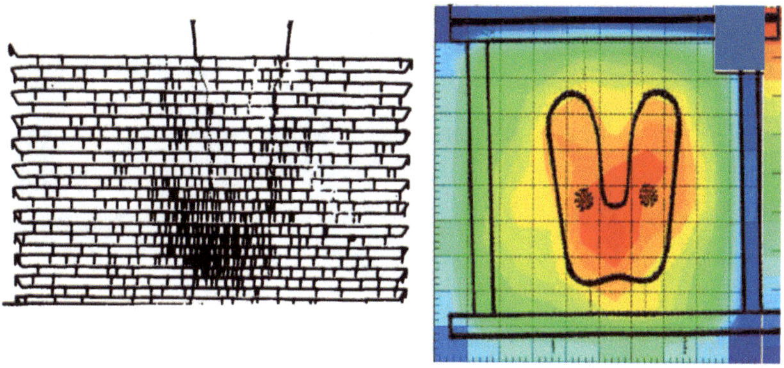

Fig. 5.7 *Left*: the first image of a patient thyroid obtained with a PMT-NaI detector by Benedict Cassen in 1950 (reprinted from [16] with permission of Elsevier Ltd). *Right*: the first image of a thyroid phantom filled with ^{131}I obtained with the recoil electron-tracking Compton camera (Reprinted from [14] with permission of Elsevier Ltd)

5.2.4 *The SPECT Ultimate Goal: Multiple Coincidences Compton Camera*

Using several Compton cameras positioned along a polygon setup around the patient, all the 360° directions are covered removing the need to rotate the camera. In addition, contrary to multi-head SPECT using mechanical collimators, the system can record all the γ-rays originating from a single decay.

In 1987, Liang et al. [17] pointed out that, using these Compton cameras in coincidence, the decay must have occurred on the intersection of several cones (Fig. 5.8). With isotopes emitting two γ-rays in cascade such as ^{111}In or ^{67} Ga, the decay event will be located on 1 or 2 closed curve lines being the intersection of the two cones (Fig. 5.8a). Using isotope-emitting three γ-rays in cascade such as ^{130}I or ^{94}Tc, the intersection of the three cones will reduce to a finite number of points, in the most favorable case only 1 point (Fig. 5.8b). Of course, increasing the coincidence multiplicity decreases the system efficiency: keeping only triple coincidences events efficiency is expected to fall to that of conventional SPECT ($\approx 10^{-4}$) [18]. However, each of such recorded events will be much more informative than in no-coincidence mode, and by only considering triple-coincidence intersections, reducing to 1 point will make the reconstruction process superfluous (the benefits of this feature have already been discussed in Sect. 5.1).

Using several recoil electron-tracking Compton cameras like the one proposed by Kabuki, all the two γ-rays coincidences will directly provide the decay position (Fig. 5.8c). To this end, the coincidence triggering has to be performed using the first detector, i.e., the Compton camera that is a fast silicon detector. A recent prototype of such Compton camera working at atmospheric pressure showed a single-photon efficiency of 1.5×10^{-4} for energies above 150 keV and for a $10 \times 10 \times 10$ cm^3 detection volume [15], equivalent to an efficiency of about

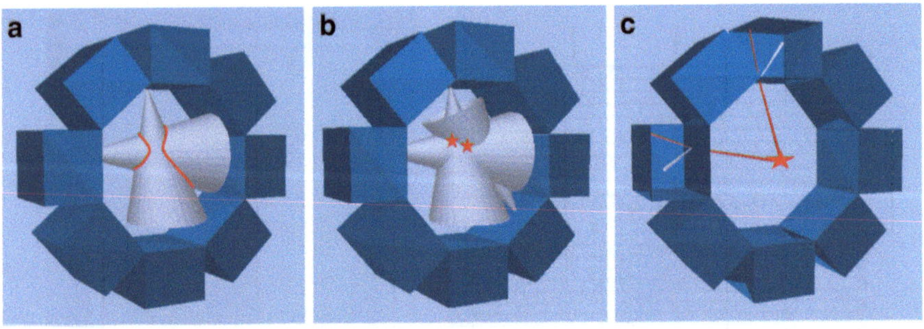

Fig. 5.8 Improved accurate position information can be obtained using an octagon of 8 Compton cameras working in coincidence. (**a**, **b**) conventional Compton camera with isotopes emitting at least two or at least three γ-rays, respectively. (**c**) Recoil electron-tracking Compton camera with isotope emitting at least two γ-rays (top and left detectors are shown open to illustrate the recoil photon and the recoil electron tracking)

0.4 % for 30 × 30 × 30 cm^3 detection volume. Simulations have shown that at higher-pressure values, this detector volume could provide an efficiency of about 2 % [14]. Surrounding the patient by such detectors, such as in Fig. 5.8, the single-photon efficiency should be around 10 %. When working in two γ-rays coincidences, the sensitivity should be similar to that of 3D PET, i.e., ≈1 %, but again without the need of using a reconstruction process. Substances labeled with ^{94}Tc(4.9h), ^{95}Tc(20h), ^{96}Tc(103h), ^{111}In(67h), ^{198}Tl(7h), ^{199}Tl(7h), and ^{67}Ga(78h) will be usable and, as a consequence, also most of the radiopharmaceuticals developed for SPECT. For isotopes emitting n γ-rays in cascade per decay, the efficiency of detecting at least two photons in coincidences is $n \times \text{eff} \times (n-1) \times \text{eff}$, where eff is the detector efficiency. Thus, for ^{96}Tc that emits at least three γ-rays of energy around 800 keV, the efficiency will thus further be improved by a factor 3 versus a radionuclide-emitting two photons. Obviously, such sensitivity estimations are based on preliminary simulations and have still to be confirmed by real prototypes.

5.3 New Technology Developments in PET

5.3.1 The PET Ultimate Goal: 25 ps TOF-PET

The concept of positron imaging was introduced by Sweet and Brownell in 1950 and is anterior to the development of the γ camera by Anger starting in 1952 [19]. A first positron image of a patient was already obtained in 1953 using a rectilinear scanner made of two opposing detectors [19]. However, the first true positron camera was constructed in the late 1960s and consisted of two Anger γ cameras. The hardware collimator is the major component limiting the sensitivity and spatial resolution of single-photon imaging system, and its replacement by the electronic collimation in positron imaging resulted in a significant improvement.

The real precursor of modern emission tomography was Kuhl and Edward's scanner [20] in 1964, which used two opposing detectors making a single sweep at angular orientations around 360°. However, the system was purely analogical and provided just a back projection, i.e., without ramp filtering, of the acquired projection lines. A major improvement was the introduction in the beginning of the 1970s of the computers that allowed inverting the Radon transform to reconstruct the 3D activity distribution that is supposed to have generated the acquired planar views.

During time, with the improvement of the computer speed, more and more sophisticated algorithms were introduced, which better modeled the acquisition process and the Poisson nature of the nuclear decays. That resulted in a continuous improvement of SPECT and PET quality. Recently, with the measurement of the time of flight (TOF) difference between the two 511 keV γ-rays, PET is on the way to get free of the computer by directly measuring the annihilation position.

Since the introduction of PET with a filtered back-projection algorithm by Ter-Pogossian et al. in 1975 [21], the time resolution improvement of the system has been a major concern to the end of getting rid of the random coincidences, which are proportional to the coincidence time window. One should consider that this random coincidence contamination in conventional PET cannot indefinitely be reduced due to the necessity of having a coincidence time window larger than the time needed for the γ-ray to travel the field of view (FOV), i.e., 2 ns for a 60 cm FOV.

The use of TOF information to prelocalize the positron annihilation was already explored in the early 1980s. The requirement of a sub-nanosecond time resolution in order to improve the image reconstruction disqualified the BGO, and many experiments were carried out on a two-arm detector setup using CsF and BaF_2 crystals (Fig. 5.9). An excellent time resolution of 106 ps was already obtained in 1989 [22]. TOF-PET using CsF and BaF_2 crystal was built, achieving a good system time resolution of about 550 ps [23, 24]. However, the low light outputs and the low stopping power of these crystals hamper the spatial resolution and the sensitivity. Timing stability of these systems was also challenging, and their overall performances did not overpass those of conventional BGO systems. Only at the end of the 1990s, the development of somewhat less fast crystals, but with a higher light yield, raised again interest in TOF-PET systems.

The lutetium oxyorthosilicate crystal family (LSO, LYSO) was introduced (Fig. 5.9), and Philips presented in 2006 the first commercial TOF-PET with a time resolution of 650 ps.

These five last years, an intense race was engaged in developing coincidence detection systems reaching the 25-ps grail (Fig. 5.9). The best time resolutions obtained up to now in two-arm detector setup using thin LSO, $LaBr_3$, and pure CsBr crystals are 190, 100, and 75 ps, respectively [25–27]. The SiPM provided better performance than the new generation of fast PMTs for applications with LSO as well as with $LaBr_3$. But for the time being, it is too early to prognosticate which of the two components will win as they are still both in development.

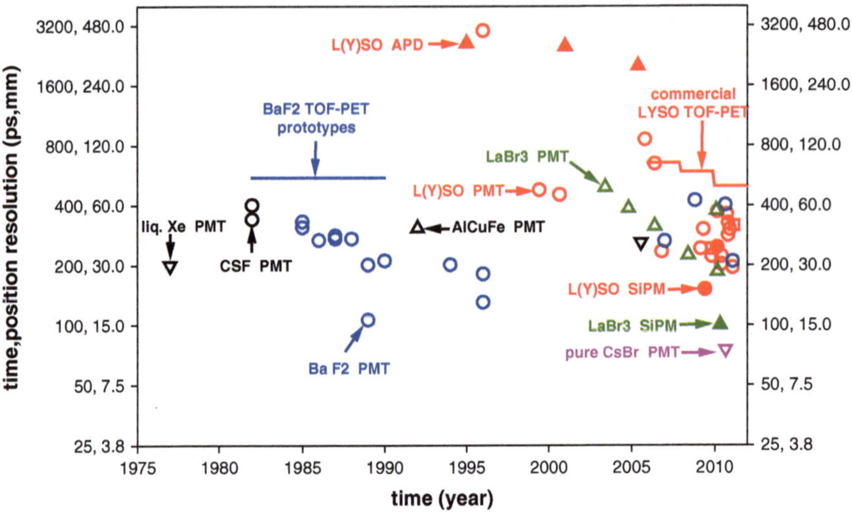

Fig. 5.9 Scientific publications reporting coincidence time resolutions in experimental two-arm detector setup. *Empty* and *filled symbols* correspond to PMTs and photodiodes, respectively. Difference in time resolution for same crystal-photodetector tandem results from differences in crystal size, photodetector model, doping level, and electronic design for signal processing. 25 ps correspond to an annihilation localization accuracy of 3.8 mm avoiding the need of a reconstruction process. Note the increasing number of publications these last 5 years

In conventional PMTs the large feature size of the dynodes and of their separation results in a spread of transit time and consequently in poor time resolution. The fast PMTs of the new generation, called micro-channel plate (MCP) PMTs, are made of a parallel array of thin microtubules containing an emissive layer for electron amplification. The pore size is typically on the order of 10 mm, and the distance from photocathode to the first amplification stage is only a few mm. One advantage of these devices is their large area, which is similar to standard PMTs.

SiPMs, in addition to be unaffected by electromagnetic fields, have the benefit to be transparent to 511 keV γ-rays. This is a particularly advantageous feature when using low stopping power crystals such as $LaBr_3$, which require a large thickness to reach a good sensitivity. A large thickness affects indeed time resolution, as the photoelectric effect may occur at different depth into the crystal. Using SiPMs, it is possible to stack several thin crystal-SiPM layers to solve the issue.

Up to now, Shibuya et al. [27] achieved the best time resolution using a pure CsBr crystal, i.e., without doping impurities. Doping is used to increase the light yield of crystals but often results in a dramatic decrease of the speed performance. Recent progress in photon detectors and electronics has considerably reduced the requirement for scintillation materials in terms of the amount of light output to be obtained, and some pure crystals are worth reconsidering as candidates for ultrafast scintillators. Considering the photons loss and the photodetector efficiency, about 500 photons at 511 keV are needed to achieve an acceptable energy resolution of 25 %. Shibuya et al. pointed out that not all the emitted photons need to be in a single fast light component: an output of 1,000–2,000 photons/MeV, of which only one-tenth belonging to a very fast component with a decay time constant less than 50 ps, is sufficient to achieve good energy resolution and timing.

State-of-the-art time resolutions shown in Fig. 5.9 were obtained using two-arm detectors linked to a very straightforward coincidence electronics device. There are several additional features that alter the time resolution in a commercial PET system. More complex and longer electronics circuit is needed to watch coincidence occurrence between all possible detector pairs and to provide time stamping. Utilization of PMTs requires assembly of crystal in module. Light reflection on the boundaries in the array and summation of the signal from the four PMTs increase the time resolution compared to a single crystal-PMT setup. SiPMs array removed this last issue as each crystal pixel signal is individually processed.

5.3.2 Dual-Isotope PET

As all γ-rays originating from the positron annihilation have their energy very close to that of the electron at rest (511 keV), PET systems cannot distinguish between two different isotopes which are both pure positron emitters. This fact makes dual isotopes acquisitions like the ones performed in SPECT impossible. Andreyev and Celler [28] proposed an elegant way to remove this limitation. The idea is to use a "pure" positron emitter (such as ^{18}F, ^{11}C, ^{15}O) simultaneously with a "dirty" positron emitter, i.e., an isotope which additionally emits prompt single γ-rays (such as ^{86}Y, ^{124}I, ^{22}Na, ^{60}Cu). A third γ-ray detected in coincidence with the two 511 keV γ-rays is used to identify the decay of the "dirty" isotope.

Obviously the "pure" positron emitter acquisition will be corrupted by "dirty" decay events when no prompt single γ-ray is detected as a consequence of limited system sensitivity or of tissue absorption in patient. However, the probability of such scenario can be computed or experimentally measured in phantoms, and a corresponding fraction of the "dirty" isotope sinogram can be removed from the "pure" isotope sinogram before reconstruction. It should be remembered that such crossover contamination is also present in dual-isotope SPECT due to the scattering down of the highest energy γ-rays into the lowest energy acquisition window.

A clinical application of dual-isotope PET could be imaging of tumor hypoxia using ^{60}Cu-diacetyl-di(N^4-methylthiosemicarbazone) (^{60}Cu-ATSM) combined with the standard study of glucose metabolism using ^{18}F-FDG. Another application would be simultaneous imaging of myocardial hypoxia and perfusion using ^{60}Cu-PTSM and ^{11}C-acetate. Performing the two acquisitions simultaneously will be more comfortable for the patient, will reduce the patient irradiation by requiring only one X-ray CT, will be more costly efficient, and will avoid imprecision due to possible metabolic changes between two sequential scans.

The main current limitations to the use of this technique are that the "dirty" isotope reconstruction has to include appropriate spurious coincidence correction [29] and also that companies have to upgrade their acquisition software in order to record triple, or multiple, coincidences.

5.4 New Clinical Applications: In Line with Therapy

5.4.1 Real-Time Microsphere Deposition Tracking in Liver SIRT by Pinhole Bremsstrahlung SPECT

Selective internal radiation therapy (SIRT) using ^{90}Y-labeled microspheres is a rapidly emerging treatment of unresectable chemorefractory primary and metastatic liver tumors. The success of such therapeutic approach depends on (1) the expertise of the interventional radiologist to selectively catheterize the appropriate branch of artery, (2) the selection of patients with limited tumor burden, and (3) the determination of the maximal activity which can be safely injected into the patient's body. This determination is not achievable by angiography and is usually performed using empirical formulas, such as the partition model [30]. Pre-therapy SPECT using ^{99m}Tc-labeled macroaggregates (^{99m}Tc-MAA) is mainly intended to rule out patients who display a liver-to-lung shunt in excess of 20 %. Indeed, compared to ^{99m}Tc-MAA, the higher number of ^{90}Y-microspheres injected during the therapeutic procedure leads to a more pronounced embolic effect. This can significantly alter the arterial flow distribution, resulting in a distribution of the ^{90}Y-microspheres different than that of the ^{99m}Tc-MAA [31, 32]. Several methods are already used to clinically assess the microsphere deposition after SIRT and check that the procedure has been performed as expected; however, a methodology able to quickly image the microsphere deposition during the liver SIRT would be desirable.

Gupta et al. [33] showed the feasibility of real-time visualization of iron-labeled microspheres delivery during liver SIRT in rabbits using magnetic resonance. In this paper, cosigned by R. Salem, the authors concluded: "*Although quantitative in vivo estimation of microsphere biodistribution may prove technically challenging, the clinical effect could be enormous, thus permitting dose optimization to maximize tumor kill while limiting toxic effects on normal liver tissues.*" However, human liver SIRT appears quite incompatible with MR: the X-ray angiographic imager will difficultly be implemented around the MR table, and the long duration of liver SIRT, that can overpass 2 h for challenging arterial trees, is not supportable by the MR agenda.

The methodologies already in use for post-SIRT assessment of microsphere deposition include the use of parallel-hole bremsstrahlung SPECT [29] or the PET imaging of the internal pair electron-positron emission of ^{90}Y labeled to the microspheres [34]. Unfortunately, parallel-hole bremsstrahlung SPECT is corrupted by high-energy X-rays scattered by the collimator septa down to the acquisition energy window. Although correction techniques have been proposed, the spatial resolution and the quantification accuracy are still unsatisfactory. The

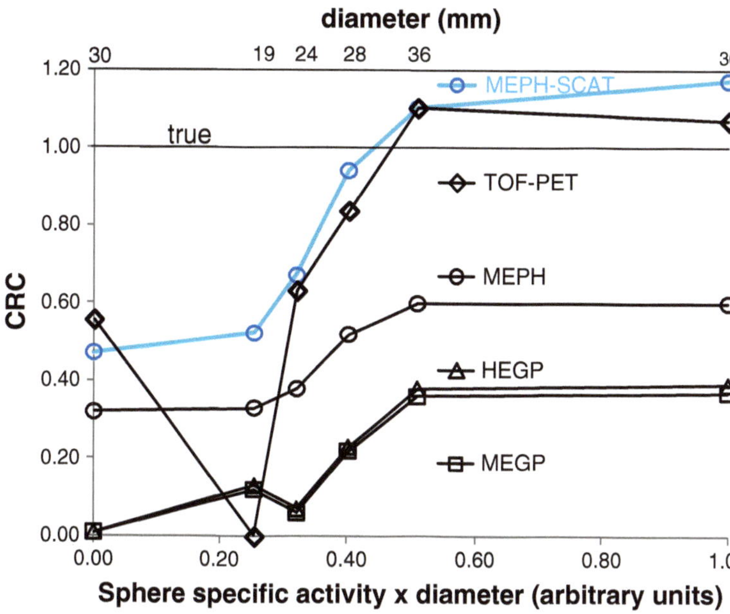

Fig. 5.10 Contrast recovery coefficient (CRC) as a function of the actual sphere-specific activity times the sphere diameter. The true CRC is that obtained with the actual activity ratio. Reprinted from [35]

internal pair electron-positron emission in ^{90}Y occurs only in 32 out of one million decays and thus requires long PET acquisition times.

Recently, it was shown that pinhole collimators are much better adapted to bremsstrahlung SPECT than parallel-hole collimators [35]. Comparison between medium-energy parallel-hole (MEGP) and medium-energy pinhole (MEPH) bremsstrahlung SPECT and TOF-PET was performed using an abdominal-shaped phantom (5 l, 0.3 GBq total ^{90}Y activity) containing hot and cold spheres: 36, 36, 28, 24, 19, and 30 mm diameter having a relative activity ratio to the background of 7, 3.5, 3.5, 3.5, 3.5, and 0, respectively. With scatter correction for TOF-PET, but not for SPECT, TOF-PET provided the best contrast for the 36-, 30-, 28-, and 24-mm diameter spheres, while MEPH SPECT gave the best visualization of the 19-mm diameter sphere (Fig. 5.10). MEPH provided significantly better contrast recovery coefficients (CRC) than MEGP. Except for the cold sphere and for the hottest one, MEPH with scatter correction gave CRC values closer to 1 than TOF-PET did. In particular, it should be noted that for the smallest spheres, TOF-PET is hampered by its lower sensitivity and therefore unable to visualize them.

In addition, ten helical MEPH SPECTs of a realistic liver-SIRT phantom (1.4 GBq ^{90}Y in 800-ml total targeted liver, tumor-/liver-specific activity = 4) were also acquired. As helical SPECT acquisition is not yet implemented in commercial γ cameras, the detector and bed motions were achieved by hands resulting in a total acquisition time of 3 h, making a trial on patients impossible. The helical MEPH SPECTs provided accurate and reproducible activity estimation

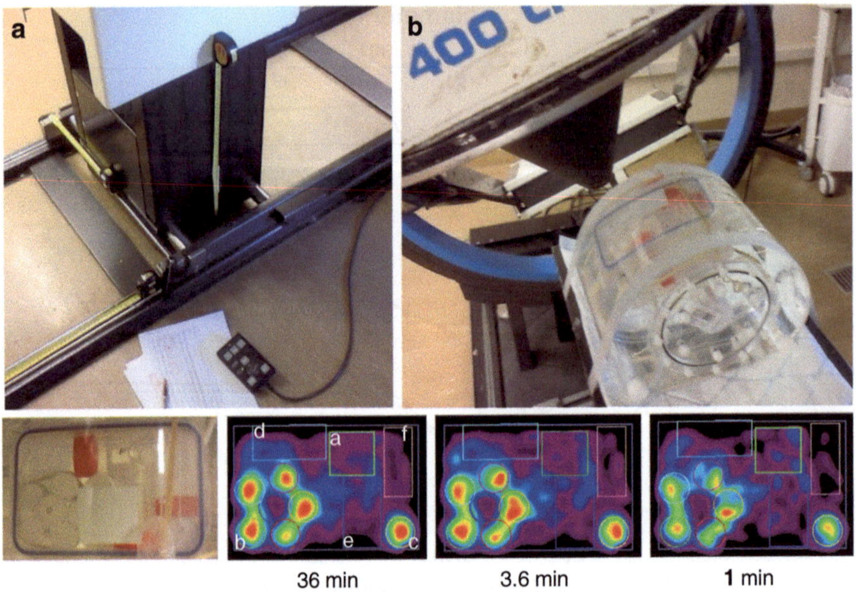

Fig. 5.11 Liver-SIRT phantom acquisition and reconstruction. (**A**) bed holder and the three tape measures. (**B**) Note the counterweight lever system that does not allow pure radial motion. *Bottom row*: picture of the liver model and reconstructed oblique slices passing though the middle of the liver model for 36-, 3.6-, and 1-min acquisition. a: VOI sample in the healthy liver. b: Necrotic tumor and c: isolated tumor with both specific activity fourfold that of the healthy liver. d, e, f: Cylinders with specific activity 0.5, 0.5, and 0.25 times that of the healthy liver, respectively. The cylinder f section is smaller than that of cylinder e and d, because cylinder f was not centered. Reprinted from [35]

of the liver-SIRT phantom for effective acquisition time going down to 1 min (relative deviation of healthy liver compartment: 10 ± 0.1 %) (Fig. 5.11).

These results support the implementation of helical pinhole bremsstrahlung SPECT acquisition in the catheterization room. The whole helical orbit of the multi-pinhole camera could be performed by a 6-axis robot arm. In home position, the system will leave free the space around the catheterization table (Fig. 5.12). No special development is required for the combined use of gamma camera and robot; only a synchronization clap is required between the gamma acquisition and the motion driver software. Such kinds of robots are already used in radiation therapy [36] or assisted surgery. State-of-the-art informatics systems controlling the robot motions are reliable and efficiently prevent any harm to the patient.

5.4.2 In-line PET in Hadron Therapy

Hadron therapy is a rapidly emerging new technique (Fig. 5.13) for single tumor ablation. Similarly to internal radiotherapy, hadron therapy uses a vector to

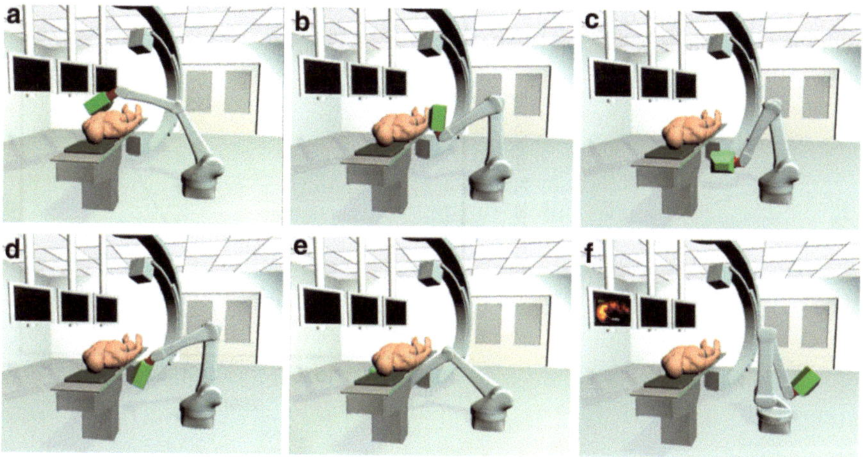

Fig. 5.12 Artistic sketch of the implementation in the catheterization room of a pinhole bremsstrahlung SPECT allowing microsphere deposition tracking during liver SIRT in order to optimize the embolization procedure. The noncircular helical SPECT acquisition motion is performed by a 6-axis arm robot. In home position (**f**), the system leaves free the space around the patient. Note in (**c**) the flipping of the detector to face the collimator to the back of the patient in view of the bottom orbit (**d**, **e**). Reprinted from [35] (mpeg animation in open access)

transport the ionizing energy into the tumor in order to limit irradiation of the surrounding tissues. The major difference with internal radiotherapy is that the vector transporting the ionizing energy in hadron therapy is the hadron itself! This apparent paradox disappears in the light of the special kinetics properties of heavy charged particles traveling through matter. High-energy photons (X- or γ-rays) and light charged particles (electrons) spread all their energy along their whole path. On the contrary, heavy charged particles, such as protons or ions, deliver most of their energy at the end of their paths as discovered by William Bragg in 1903 (Fig. 5.14) and with no energy deposition beyond that range. The depth of the Bragg peak only depends on the particle energy and on the chemical composition of the tissue. In principle, knowing the tissue composition, it is possible to plan out the hadron energy and accurately "paint" the tumor. This is a significant benefit versus internal radiotherapy where the radio compound can be taken up in a difficultly controllable way by some healthy tissues.

However, several limitations hamper this ideal scenario. Currently, the therapy planning is based on CT scan requiring a rescaling of the Hounsfield values into hadron stopping power. Unfortunately, the hadron stopping power does not exactly depend on the Z density in the same way as X-rays do, resulting in some inaccuracies in the planning. Patient repositioning can also introduce some deviations from the planned field, especially in tumors located into soft tissue, such as the neck, that can also move on their own. These limitations support the development of techniques to assess in real time the location of the energy deposition. By this way, the hadron energy could be tuned to get the Bragg peak at the right place.

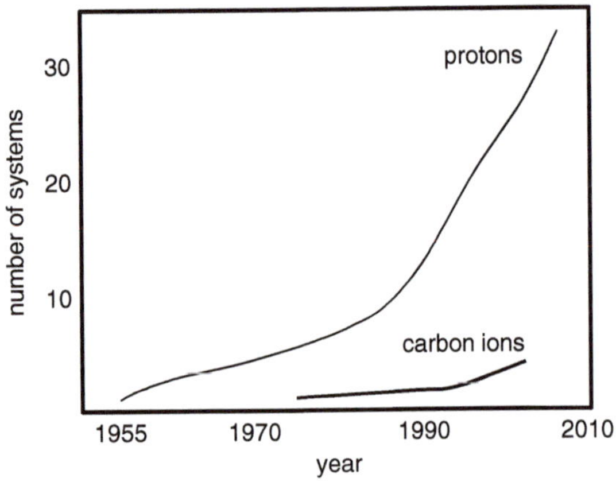

Fig. 5.13 Evolution of hadron therapy facilities with time

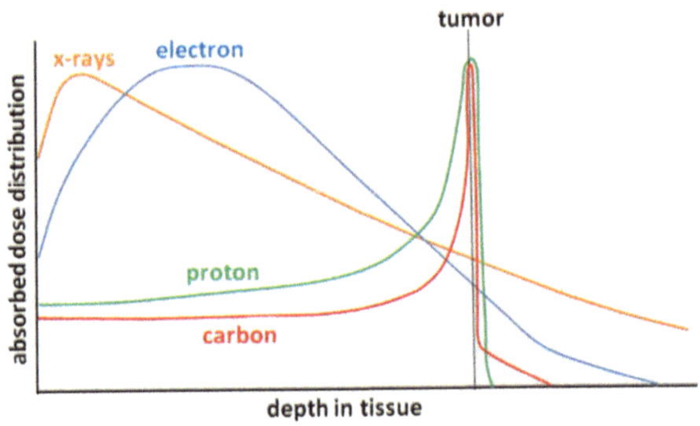

Fig. 5.14 Typical depth absorbed dose curves for photons, electron, protons, and carbon

During hadron therapy, short-lived positron emitters, mainly ^{15}O (122 s), ^{11}C (20 min), and ^{13}N (10 min), are produced in a very low abundance in the patient's body by means of nuclear fragment reactions induced by the incoming hadrons. Since 15 years these reaction products have been imaged using commercial PET systems [37–39]. However, the transportation of the patient between the hadron facility to the PET room takes several tens of minutes. The activity reduction by physical decay during the patient transportation makes this method impracticable to tune the beam parameters at the beginning of the irradiation. In addition, biological washout may also displace the activity during the patient transportation, introducing imprecision in the Bragg peak localization.

To overcome these limitations, dedicated in-line PET systems are being developed and already in clinical evaluation at the Gesellschaft für Schwerionenforschung (GSI) in Darmstadt, Germany [40], and at HIMAC [41] and Kashiwa [42, 43], in Japan and in Pisa, Italy [44]. To be implemented in a commercial hadron beam system, such PET scanners use partial ring detector. In addition to reduce the system sensitivity, this feature requires a sufficient rotation range around the patient to avoid reconstruction artifacts and anisotropic spatial resolution. Incorporating detector rotation makes the scanner gantry more complex and expensive and

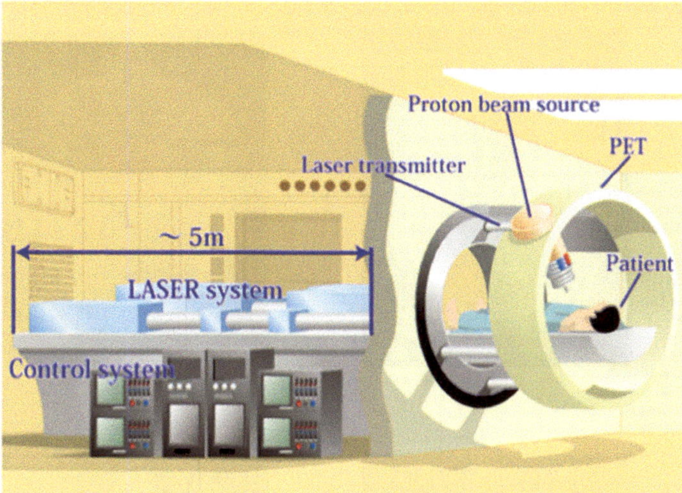

Fig. 5.15 Artistic rendition intended to emphasize potential PET application of laser-driven proton beam. The figure omits many key technical components and may have little resemblance to what a laser-driven ion beam radiotherapy facility will look like. Courtesy of Dr. Paul Bolton (Japan Atomic Energy Agency)

increases the scanning time. As a result, these PET systems cannot be used to tune the hadron beam during therapy and are mainly used as a post-therapy quality control. Alternately, there has been a recent proposal of an in situ scanner with two complete detector rings with a gap in between to accommodate the beam line [45].

Several studies have proven the use of TOF information for reducing the artifacts which result from partial ring geometry [46–49]: two-third ring scanner with 600 ps timing resolution (already commercially achieved) still provides accurate images for proton dose monitoring. By reducing the reconstruction noise with improved time resolutions, the hadron beam optimization could start sooner after the beam lightening on. Use of more compact hadron beam system such as laser-driven hadron production jointly with a direct design of the beam and PET system integration should still allow using more complete ring detection (Fig. 5.15).

There is significant difference between proton and carbon therapy regarding the spatial relationship between absorbed dose and positron emission [50]. Due to the energy dependence of the reaction cross sections, the inelastic scattering nuclear reactions tend to occur at higher proton energies than at the proton energies associated with the Bragg peak. Thus, at the Bragg peak, the concentration of the positron emitters rapidly decreases because energy deposition happens through other interactions, not inelastic collisions. As a result, the image obtained from a PET scan after proton therapy shows activity in front of the dose deposition (Fig. 5.16). On the contrary, in therapy using the heavier carbon ions, the fragmented products generate a peak of activity very close to the actual Bragg peak. The PET image thus shows the location of the Bragg peak. However, fine

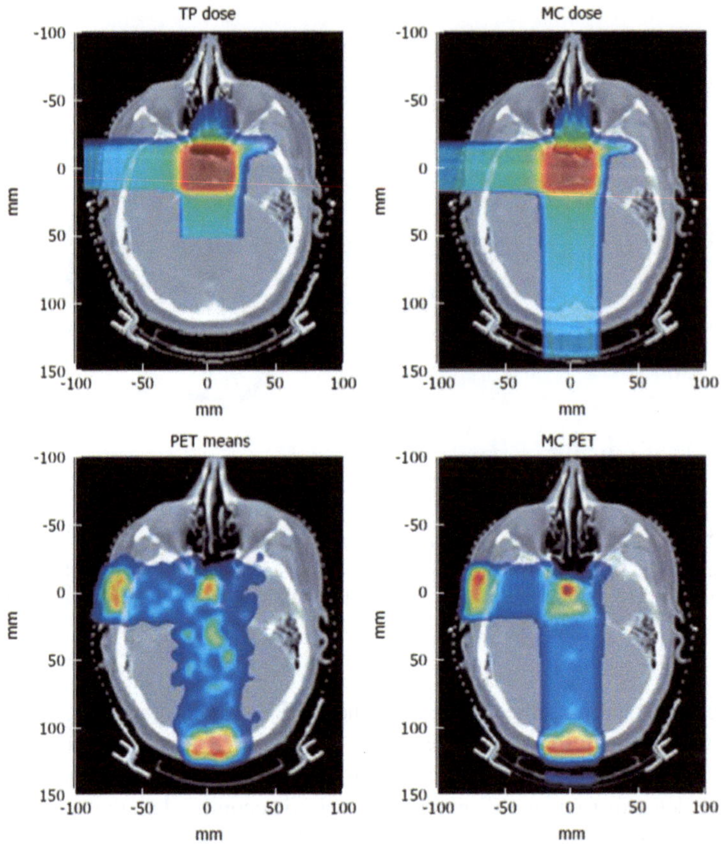

Fig. 5.16 *Top*: treatment plan (TP) and Monte Carlo (MC) dose for a patient with pituitary adenoma receiving two orthogonal fields. *Bottom*: measured and Monte Carlo PET images. Delay times from the beginning of imaging were about 26 and 18 min from the end of the first and second field applications, respectively. Range of color wash is from *blue* (minimum) to *red* (maximum) (Reprinted from [51] with permission from Elsevier Ltd)

dose delivery optimization as needed for tumor hypoxia painting will require the development of an accurate computational modeling of the spatial relationship between positron activity and dose delivery.

5.4.3 In-line Proton Computed Tomography in Hadron Therapy

Investigation on proton computed tomography (PCT) started 30 years ago and appeared to be a good alternative to X-ray CT scan in terms of more accurate reconstruction of electron density and lower dose exposure to the patient [52, 53]. However, the need of a dedicated proton accelerator made PCT dramatically more expensive than X-ray CT scan and kept PCT out of commercial developments. As PCT is directly related with proton stopping power of the patient tissue, it can in principle remove errors coming from the Hounsfield rescaling method [54]. The use of the proton accelerator already included in the hadron therapy system should improve the therapy beam planning at reasonable cost. As proton stopping power is also sensitive to the tissue chemical composition, PCT should allow localizing the tumor. In the ideal scenario, tumor localization and therapy planning by PCT should be performed just before starting the hadron therapy without moving the patient, thus solving tumor positioning issues.

The energy straggling that hampers the electron density assessment and the multiple Coulomb scattering that alters spatial resolution are still open problems in proton stopping power determination in PCT [55, 56]. A promising way to overcome these problems is to individually track each proton by measuring its direction before and after its travel inside the patient body as well as its energy loss during this travel. Monte Carlo simulations have been used in order to derive and to evaluate analytical formulation of the most likely path as a function of the in-and-out parameters of the proton [57]. Afterwards, this most likely path derived for each recorded proton is used in an iterative reconstruction process [58].

Although PCT is a transmission tomography such as X-ray CT in radiology, PCT shares important specific similarities with nuclear medicine emission tomography that are absent in X-ray CT and that strongly condition the acquisition and reconstruction process: the requirement to individually acquire each event such as in TOF-PET, the low count rate inducing noise issue in the reconstruction process, and the large fraction of nonstraight path in the patient body requiring iterative reconstruction that include sophisticated path modeling.

Modern design to measure proton in-and-out parameters consists of two trackers, also called telescopes, placed on both side of the patient body with an additional calorimeter placed at the end of the proton beam that measures the proton residual energy [59] (Fig. 5.17).

Each tracker, similar to those used in the Compton camera (Sect. 5.2.2), consists of two orthogonally oriented single-sided silicon microstrip sensors measuring the two coordinates of the particle crossing point. The sensors thickness is chosen to have a good compromise between energy loss and detection efficiency. The calorimeter is made of a crystal with good energy resolution combined with a fast response to allow individual proton energy measurement in a high-rate proton beam. A first small FOV PCT prototype has been built at the Loma Linda University Medical Center (LLUMC) California by Bashkirov et al. [60], and a reconstruction of an acrylic phantom has been carried out (Fig. 5.18). This team is currently constructing a full-scale PCT prototype large enough to cover a patient head.

A major challenge for a clinical therapy planning based on PCT is the reconstruction speed that has to be fast enough to perform the reconstruction with the patient not moving on the therapy bed. The ART algorithm, using the principle of the most likely path, sequentially updates the image for each proton history, thus representing a serial process. Due to the high number of recorded proton histories

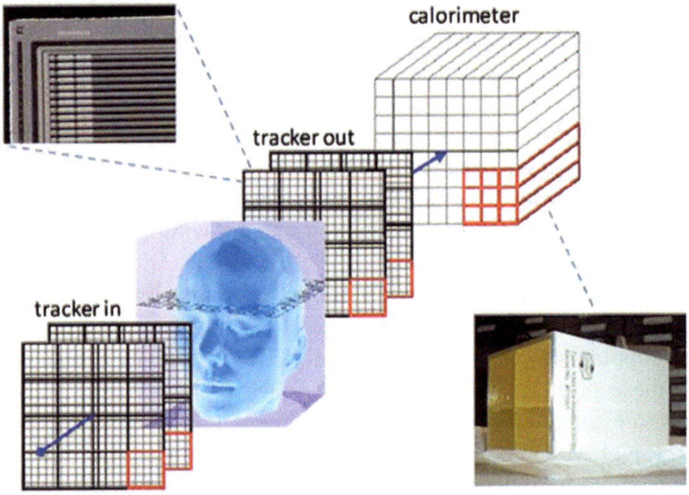

Fig. 5.17 Schematic representation of the detector setup of a suitable PCT system. Trackers are made of two orthogonal silicon microstrip detectors. The calorimeter is made of a fast and good energy resolution crystal. Reprinted from [59] with permission of IEEE

needed to have a sufficient accuracy (about 10^8), reconstruction currently requires several hours on the fastest processing units. Another algorithm, based on parallelizable diagonally relaxed orthogonal projections (DROP), is currently into development to be implemented on graphical processing units (GFU) [58]. Preliminary results suggest that accurate reconstruction could be performed in less than 10 min.

When a working full-scale PCT prototype will be achieved, the complex issue of integration will arise. Indeed, even if PCT provides an accurate therapy beam planning, in-line PET will still be needed in order to check that nothing went wrong in the planning. Integrating these two systems around the hadron beam will likely require complex gantry arrangements to swap between PCT measurement of the hadron stopping power and of the tumor position and therapy.

5.5 New Preclinical Application: Wearable PET

Brain imaging in neuroscience is hampered by the need to avoid head motion. In human, this restricts to study the brain metabolism during pure intellectual activity. In animal, this required to anaesthetize or to fully immobilize the animal: the first solution precludes brain imaging in response to sensorial stimulations, and the second disturbs the brain function as a result of the important stress imposed to the animal. Given the transient nature of the neurologic processes, poststimulation or post-behavioral brain imaging results in information loss.

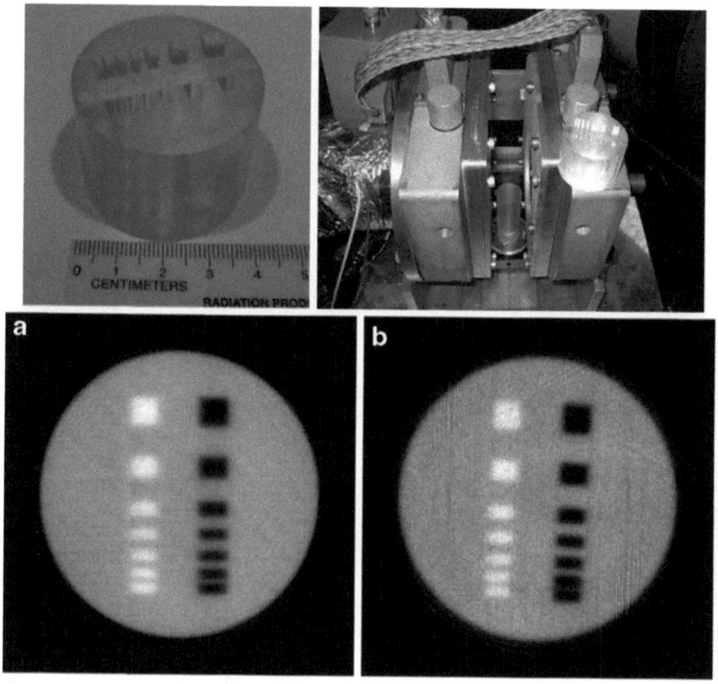

Fig. 5.18 Acrylic phantom, where one row is filled with water and the other with air. PCT detector setup. Transversal cuts of the 3D phantom image reconstructed from (**a**) simulated and (**b**) experimental data. Reprinted from [60] with permission of American Institute of Physics

To solve this issue, Paul Vaska et al. [61] started in 2001 the development of a "wearable" rat PET scanner vividly called RatCAP (rat conscious animal PET). The RatCAP scanner is a miniature PET system including a complete LYSO-APD ring, weighing only 250 g, with an inner diameter of 38 mm, outer diameter of 80 mm, and axial extent of 25 mm (Fig. 5.19c). The LYSO and APD are both in a 4×8 array geometry with a 2.2-mm square cross-sectional forming a detector block. The small LYSO thickness (5 mm) reduces the parallax issues. A front-end microchip conditions the analog signals of the block, stamps the event time with a sub-nanosecond resolution, and serializes the data into a single digital data output line. Each individual gamma-ray event is recorded in list mode; afterwards, prompt and random coincidences are built by software. The spatial resolution is better than 2 mm across the field of view, and the coincidence sensitivity is 0.76 % at the center [62].

The RatCAP is positioned between the eyes and ears of the rat (Fig. 5.19b), allowing a normal forward-facing posture while brain imaging. To allow the rat easily moving without being hampered by the PET weight, the RatCAP is suspended on long springs fixed on a rotation stage (Fig. 5.19a). Rats move freely and appeared to adapt well to the device as shown by the corticosterone blood level

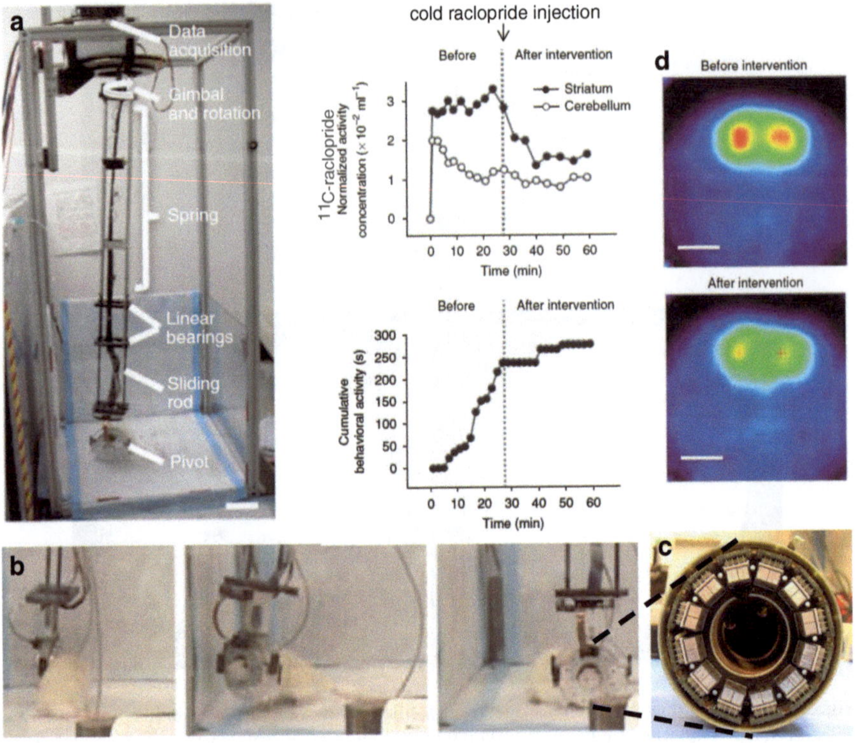

Fig. 5.19 Miniature wearable RatCAP (rat conscious animal PET). (**a**) PET weight counterbalancing system allowing free rat motions. (**b**) Rat wearing the RatCAP and freely behaving. (**c**) Cross section of the RatCAP. (**d**) Behavioral dynamic brain study. Reprinted from [62] with permission of Nature Publishing Group

that rapidly returns to normal level after an initial increase when the RatCAP is fixed to the head.

Recently, a 1-h brain dynamic ^{11}C-raclopride imaging of a rat freely behaving was performed [62]. After thirty minutes, unlabeled raclopride was given to the rat, which changed its behavioral activity, reduced the uptake in the striatum by more than 50 %, and continued to block the uptake of ^{11}C-raclopride for the remainder of the scan (Fig. 5.19d). This showed that direct correlation of PET and behavioral data under natural conditions as well as under drug could be studied. Furthermore, RatCAP proved to be fully compatible with MR system, and simultaneous PET-MR images of brain and heart in anaesthetized rats and mice were acquired [63].

Such wearable PET could still be more interesting in human neuroscience. Combined with a pump ensuring a constant ^{18}F-FDG blood level, autoregulated by simply measuring the count rate at the wrist, human brain metabolism in volunteers, or in patients, could be studied while performing complex activities, such as playing music instrument, walking, keeping equilibrium on a cable, finding the way in a labyrinth, laughing, and writing. Blood probes measuring glycemic level on the fly are nowadays available in order to correct the ^{18}F-FDG brain uptake for the natural glucose competition. The cap could cover almost half the total solid angle around the brain, and the LYSO array could be thicker than for the RatCAP, providing a high sensitivity. Due to the good FDG uptake of the brain in working state, low total activity could be used. Given the wide range of PET radiotracers to probe various aspects of brain function, this multidimensional approach can have considerable potential for yielding new insights neurology.

5.6 Conclusion

This walk in nuclear medicine imaging showed a very active research domain. Techniques, such as Compton SPECT camera, TOF-PET, and PCT, already proposed more than 30 years ago, begin to become real, boosted by the development of new detector components: avalanche photodiode, silicon microstrip detector, and fast scintillation crystal. The exciting grail quest of a system directly giving the decay position, event by event, appears now possible, both in SPECT and PET. More immediately, the constant improvement of time resolution in TOF-PET will continuously reduce noise and artifacts in reconstructed image and will soon open TOF-PET to diagnostics using low branching positron emitters, such as already made for ^{90}Y but still currently limited to therapeutic activities. Lastly, nuclear medicine imaging will become an active and essential partner during some radiotherapies, namely, liver-SIRT and hadron therapy.

References

1. Karp JS, Surti S, Daube-Witherspoon ME, Muehllehner G (2008) Benefit of time-of-flight in PET: experimental and clinical results. J Nucl Med 49:462–470
2. Kadrmas DJ, Casey ME, Conti M et al (2009) Impact of time-of-flight on PET tumor detection. J Nucl Med 50:1315–1323
3. Campbell W (2009) Sensitivity improvements in TOF PET given 20 ps resolution. Spring 2009 P521A – experimental techniques in particle physics. University of Victoria, pp 1–6
4. Iams H, Salzberg B (1935) The secondary emission phototube. IEEE Proc IRE 23(1):55–64, http://dx.doi.org/10.1109/JRPROC.1935.227243
5. Choi Y (2008) Improved MR-PET components. MR-PET Workshop. Forschungszentrum Jülich, Germany, 27–28 Oct. http://www.iss.infn.it/topem/PET-MRI/slides-julich-workshop/Session4-Choi.pdf
6. Slomka PJ, Patton JA, Berman DS, Germano G (2009) Advances in technical aspects of myocardial perfusion SPECT imaging. J Nucl Cardiol 16(2):255–276
7. Todd RW, Nightingale JM, Everett DB (1974) A proposed γ camera. Nature 251:132–134
8. Smith B, Matthews KL, Lackie A et al (2006) An electronically-collimated gamma-ray detector for localization of radiation sources. 2006 IEEE nuclear science symposium conference record, pp 257–263

9. Seo H, Kim CH, Park JH et al (2010) Development of double-scattering-type Compton camera with double-sided silicon strip detectors and NaI(Tl) scintillation detector. Nucl Instrum Methods Phys Res A 615(3):333–339
10. Roellinghoff F, Richard M-H, Chevallier M et al (2011) Design of a Compton camera for 3D prompt-γ imaging during ion beam therapy. Nucl Instrum Methods Phys Res A 1–4
11. Watanabe S, Takeda S, Ishikawa S et al (2007) Development of semiconductor imaging detectors for a Si/CdTe Compton camera. Nucl Instrum Methods Phys Res A 579(2):871–877
12. Clinthorne NH, Burdette D, Studen A, Honscheid K, Kagan H, Chesi E, Huh S, Lacasta C, Llosa G, Mikuz M, Smith DS, Weilhammer P (2005) Timing in silicon pad detectors for Compton cameras and high resolution PET. IEEE nuclear science symposium conference record
13. Mihailescu L, Vetter K, Burks M, Hull E, Craig W (2007) SPEIR: A Ge Compton camera. Nucl Instrum Methods Phys Res A 570(1):89–100
14. Kabuki S, Hattori K, Kohara R et al (2007) Development of electron tracking Compton camera using micro pixel gas chamber for medical imaging. Nucl Instrum Methods Phys Res A 580 (2):1031–1035
15. Takada A, Kubo H, Nishimur H et al. Observation of diffuse cosmic and atmospheric gamma rays at balloon altitudes with an electron-tracking Compton camera. http://arxiv.org/PS_cache/arxiv/pdf/1103/1103.3436v2.pdf
16. Blahd WH (1996) Ben Cassen and the development of the rectilinear scanner. Semin Nucl Med 26(3):165–170
17. Liang Z, Hart H, Schoenfeld A (1987) Triple gamma coincidence tomographic imaging without image processing. IEEE Conf Eng Med Biol 9:825–826
18. Andreyev A, Sitek A, Celler A (2010) Spatial resolution of the multiple coincidences Compton camera. Nuclear science symposium conference record (NSS/MIC), pp 2099–2103
19. Mccready VR (2000) Milestones in nuclear medicine. Eur J Nucl Med 27(Suppl):S49–S79
20. Kuhl DE, Edwards RQ (1964) Cylindrical and section radioisotope scanning of the liver and brain. Radiology 83:926–936
21. Ter Pogossian MM, Phelps ME, Hoffman EJ et al (1975) Positron emission transaxial tomograph for nuclear imaging (PETT). Radiology 114:89–98
22. Moszyńskia M, Mach H (1989) A method for picosecond lifetime measurements for neutron-rich nuclei: (2) Timing study with scintillation counters. Nucl Instrum Methods Phys Res A 277(2–3):407–417
23. Lewellen TK, Bice AN, Harrison F, Pencke MD, Link JM (1988) Performance measurements of the sp3000/uw time-of-flight positron emission tomograph. IEEE Trans Nucl Sci 35(1): 665–669
24. Soussaline F, Campagnolo R, Verrey B, Bendriem B, Bouvier A, Lecomte JL, Comar D (1984) Physical characterization of a time-of-flight positron emission tomography system for whole-body quantitative studies. Annual meeting of the Society of Nuclear Medicine, Los Angeles, CA, 5 Jun
25. Hui L, Yundong S, Kai Z, Jingbiao P, Zhu W (2011) A simplified digital positron lifetime spectrometer based on a fast digital oscilloscope. Nucl Instrum Methods Phys Res A 625: 29–34
26. Schaart DR, Seifert S, Vinke R et al (2010) LaBr 3: Ce and SiPMs for time-of-flight PET: achieving 100 ps coincidence resolving time. Phys Med Biol 55:N179–N189
27. Shibuya K, Saito H, Koshimizu M, Asai K (2010) Outstanding timing resolution of pure CsBr scintillators for coincidence measurements of positron annihilation radiation. Appl Phys Express 3. doi:10.1143/APEX.3.086401
28. Andreyev A, Celler A (2011) Dual-isotope PET using positron-gamma emitters. Phys Med Biol 56(14):4539–4556
29. Walrand S, Flux GD, Konijnenberg MW et al (2011) Dosimetry of yttrium-labelled radiopharmaceuticals for internal therapy: 86 Y or 90 Y imaging? Eur J Nucl Med Mol Imaging 38(Suppl 1): S57–S68

30. Kennedy A, Nag S, Salem R, Murthy R, McEwan AJ, Nutting C, Benson A, Espat J, Bilbao JI, Sharma RA, Thomas JP, Coldwell D (2007) Recommendations for radioembolization of hepatic malignancies using yttrium-90 microsphere brachytherapy: a consensus panel report from the radioembolization brachytherapy oncology consortium. Int J Radiat Oncol Biol Phys 68(1):13–23
31. Lambert B, Mertens J, Sturm EJ, Stienaers S, Defreyne L, D'Asseler Y (2010) 99mTc-labelled macroaggregated albumin (MAA) scintigraphy for planning treatment with 90Y microspheres. Eur J Nucl Med Mol Imaging 37(12):2328–2333
32. Bilbao JI, Reiser MF (2008) Liver radioembolization with 90Y microspheres. In: Baert AL (ed) Medical radiology – diagnostic imaging. Springer, Berlin, p 86
33. Gupta T, Virmani S et al (2008) MR tracking of iron-labeled glass radioembolization microspheres during transcatheter delivery to rabbit VX2 livers tumors: feasibility study. Radiology 249(3):845–854
34. Lhommel R, Goffette P, Van den Eynde M, Jamar F, Pauwels S, Bilbao JI, Walrand S (2009) Yttrium-90 TOF PET scan demonstrates high-resolution biodistribution after liver SIRT. Eur J Nucl Med Mol Imaging. doi:10.1007/s00259-009-1210-1
35. Walrand S, Hesse M, Demonceau G, Pauwels S, Jamar F (2011) Yttrium-90 labeled microspheres tracking during liver selective internal radiotherapy by bremsstrahlung pinhole SPECT: feasibility study and evaluation in an abdominal phantom. EJNMMI Res 1:32. http://www.ejnmmires.com
36. Kraus-Tiefenbacher U, Scheda A, Steil V, Hermann B, Kehrer T, Bauer L, Melchert F, Wenz F (2005) Intraoperative radiotherapy (IORT) for breast cancer using the Intrabeam system. Tumori 91(4):339–345
37. Oelfke U, Lam GKY, Atkins MS (1996) Proton dose monitoring with PET: quantitative studies in Lucite. Phys Med Biol 41:177–196
38. Pawelke J, Enghardt W, Haberer T, Hasch BG, Hinz R, Kramer M, Lauckner E, Sobiella M (1997) In-beam PET imaging for the control of heavy-ion tumour therapy. IEEE Trans Nucl Sci 44:1492–1498
39. Litzenberg DW, Roberts DA, Lee MY, Pham K, Molen AMV, Ronningen R, Becchetti FD (1999) On-line monitoring of radiotherapy beams: experimental results with proton beams. Med Phys 26:992–1006
40. Enghardt W, Debus J, Haberer T, Hasch BG, Hinz R, Jakel O, Kramer M, Lauckner K, Pawelke J, Ponisch F (1999) Positron emission tomography for quality assurance of cancer therapy with light ion beams. Nucl Phys 654:1047c–1050c
41. Iseki Y et al (2003) Positron camera for range verification of heavy-ion radiotherapy. Nucl Instrum Methods Phys Res A 515:840–849
42. Nishio T, Miyatake A, Ogino T, Nakagawa K, Saijo N, Esumi H (2010) The development and clinical use of a beam on-line PET system mounted on a rotating gantry port in proton therapy. Int J Radiat Oncol Biol Phys 6:277–286
43. Nishio T, Ogino T, Nomura K, Uchida H (2006) Dose-volume delivery guided proton therapy using beam on-line PET system. Med Phys 33:4190–4197
44. Attanasi F et al (2008) Preliminary results of an in-beam PET prototype for proton therapy. Nucl Instrum Methods Phys Res A 591:296–299
45. Yamaya T, Inaniwa T, Minohara S, Yoshida E, Inadama N, Nishikido FKS, Lam CF, Murayama H (2008) A proposal of an open PET geometry. Phys Med Biol 53:757–773
46. Vandenberghe S, Lemahieu I (2007) System characteristics of simulated limited angle TOF PET. Nucl Instrum Methods Phys Res A 71:480–483
47. Surti S, Karp JS (2008) Design considerations for a limited angle, dedicated breast, TOF PET scanner. Phys Med Biol 53:2911–2921
48. Crespo P, Shakirin G, Fiedler F, Enghardt W, Wagner A (2007) Direct time-of-flight for quantitative, real-time in-beam PET: a concept and feasibility study. Phys Med Biol 52: 6795–6811

49. Surti S, Zou W, Daube-Witherspoon ME, McDonough J, Karp JS (2011) Design study of an in situ PET scanner for use in proton beam therapy. Phys Med Biol 56:2667–2685
50. Studenski MT, Xiao Y (2010) Proton therapy dosimetry using positron emission tomography. World J Radiol 2(4):135–142
51. Parodi K, Paganetti H, Shih HA, Michaud S, Loeffler JS, DeLaney TF, Liebsch NJ, Munzenrider JE, Fischman AJ, Knopf A, Bortfeld T (2007) Patient study of in vivo verification of beam delivery and range, using positron emission tomography and computed tomography imaging after proton therapy. Int J Radiat Oncol Biol Phys 68:920–934
52. Hanson KM (1979) Proton computed tomography. IEEE Trans Nucl Sci 26(1):1635–1640
53. Koehler AM, Steward VW (1973) Proton beam radiography in tumor detection. Science 179 (4076):913–914
54. Zygmanski P, Gall KP, Rabin MSZ, Rosenthal SL (2000) The measurement of proton stopping power using proton-cone-beam computed tomography. Phys Med Biol 45:511–528
55. Schulte R, Bashkirov B, Klock ML, Li T, Wroe A, Evseev I, Williams D, Satogata T (2004) Density resolution of proton computed tomography. Med Phys 2:1035–1046
56. Schulte R, Penfold S, Tafas J, Schubert K (2008) A maximum likelihood proton path formalism for application in proton computed tomography. Med Phys 35:4849–4856
57. Erdelyi B (2009) A comprehensive study of the most likely path formalism for proton-computed tomography. Phys Med Biol 54(20):6095–6122
58. Penfold SN, Bashkirov V, Rosenfeld AB (2009) Fast and accurate proton computed tomography image reconstruction for applications in proton therapy. Simulation 213–216
59. Petterson M, Blumenkrantz N, Feldt J et al (2006) Nuclear science symposium conference record. IEEE 4:2276–2280
60. Bashkirov V, Schulte R, Coutrakon G et al (2009) Development of proton computed tomography for applications in proton therapy. AIP Conf Proc 4:460–463
61. Vaska P, Schlyer DJ, Woody CL, Stoll SP, Radeka V, Volkow N (2001) Imaging the unanesthetized rat brain with PET: a feasibility study. IEEE nuclear science symposium conference record
62. Schulz D, Southekal S, Junnarkar S et al (2011) Simultaneous assessment of rodent behavior and neurochemistry using a miniature positron emission tomograph. Nat Methods 8(4): 347–352
63. Maramraju S, Ravindranath B, Junnarkar S et al (2009) Simultaneous MRI/PET image acquisition from an MRI compatible positron emission tomography system. Proceedings of the International Society for Magnetic Resonance in Medicine, 17th scientific meeting, Honolulu, Hawaii

Part II
Data Acquisition and Image Processing

Chapter 6
Reconstruction Algorithms and Scanning Geometries in Tomographic Imaging

Oleg Tischenko and Christoph Hoeschen

In general tomographic imaging consists of two steps: the acquisition of data and the reconstruction. Thereby the following triple problem has to be solved: choose an efficient reconstruction algorithm, identify the optimal sampling conditions imposed on measured data as required by this reconstruction algorithm, and find an efficient way of collecting such data in the practice.

In this chapter the three different classes of 2D analytic reconstruction algorithms are described (1) convolution-back projection-based algorithms, (2) algorithms based on expansion in the basis of orthogonal polynomials, and (3) algorithms based on back projecting the derivative of projections. The basic inversion formulae from each of these classes are presented along with a discussion of the corresponding optimal geometry of data supposed to be inverted with them. Finally, the original idea to extent 2D reconstruction algorithms to 3D case is considered.

6.1 Introduction

All forms of computed tomography deal with cross-sectional imaging of a human body or, more precisely, of a distribution of a specific parameter within the human body. This parameter can be either the attenuation coefficient, the concentration of a radiopharmaceutical, or any other parameter, provided that the morphology of its distribution agrees with the anatomy or functioning of the object.

In practice, data are acquired in the form of integral values of the specific parameter measured along lines crossing the object. A spatial configuration of all lines of integration is referred to as either the geometry of the data or the scanning geometry.

O. Tischenko (✉) • C. Hoeschen
Research Unit Medical Radiation Physics and Diagnostics, Helmholtz Zentrum München - German Research Center for Environmental Health, Neuherberg, Germany
e-mail: oleg.tischenko@helmholtz-muenchen.de

A. Giussani and C. Hoeschen (eds.), *Imaging in Nuclear Medicine*,
DOI 10.1007/978-3-642-31415-5_6, © Springer-Verlag Berlin Heidelberg 2013

Hence, the technique that enables tomographic imaging consists of two steps: the acquisition of data within a certain scanning geometry and the reconstruction via mathematical algorithms.

Reconstruction algorithms can roughly be divided into two groups: geometry-conditioned and geometry-unconditioned algorithms. We call the algorithm geometry conditioned if its performance depends strongly on the scanning geometry. An example of a geometry-unconditioned algorithm is the algebraic reconstruction technique (ART). This is an approach where the object function is represented by the superposition of shifted copies of a compactly supported symmetrical function (e.g., Kaiser–Bessel function studied in [1]). The unknown coefficients of this superposition are solutions of a linear algebraic system, where each distinct equation relates to a certain ray. As a rule, the solution of such a system is not unique. Additional information may be used to select one reconstruction from the family of solutions. ART-style algorithms are of advantage if (a) only few projections are available, (b) the projections are either distributed irregularly or have only been taken over a limited range of angles, or (c) the data are very noisy.

In contrast to ART algorithms, the performance of the so-called analytic algorithms strictly depends on the scanning geometry. Basic inversion formulae of these algorithms are represented by multiple integrals that have to be appropriately approximated in practical applications. Obviously, the quality of the approximation depends on the sampling conditions imposed upon the function under the integral sign; these in turn are imposed by the measuring process. Since most modern scanners acquire many projection data at regularly spaced intervals, analytic reconstruction algorithms are still preferred in the transmission tomography because they are inherently much faster than ART-like algorithms and perform well if the data are sampled adequately. In the following, three classes of such algorithms are considered, namely, convolution-back projection-based algorithms, algorithms based on expansion in basis of orthogonal polynomials, and algorithms based on the back projection of the derivative of projections. The basic inversion formulae from each of these classes will be presented along with a discussion of the corresponding optimal geometry of data supposed to be inverted with them. Finally, the basic original idea to extent 2D reconstruction algorithms to 3D case is considered.

6.2 Denotations and Theoretical Background

We denote with R the field of real numbers, and with R^2 Euclidean plane, the elements of which are vectors: $\mathbf{r} \in \mathrm{R}^2$ means that $\mathbf{r} = (x, y)$, where both x and y are real. For any two vectors $\mathbf{a}$ and $\mathbf{b}$ from R^2, their scalar product is defined by $\mathbf{a} \cdot \mathbf{b} = x_a x_b + y_a y_b$. With $\boldsymbol{\tau}_\phi$, we denote the unit vector $\boldsymbol{\tau}_\phi = (\cos\phi, \sin\phi)$. The distribution of the specific parameter within the slice through the body is described by the object function $f : \mathrm{D} \to \mathrm{R}$ supported on the disk $\mathrm{D} := \{(\mathrm{x}, \mathrm{y});\ \mathrm{x}^2 + \mathrm{y}^2 \leq \mathrm{R}^2\}$. It is supposed that $f \in \mathrm{L}_2(\mathrm{D})$, which means that $\int_{\mathrm{D}} f^2(\mathbf{r}) d\mathbf{r} < \infty,\ \mathbf{r} \in \mathrm{R}^2$. The Fourier transform of f is defined by

$$\hat{f}(\boldsymbol{\lambda}) = \int_{\mathrm{D}} f(\mathbf{r})\mathrm{e}^{-\mathrm{i}(\mathbf{r}\cdot\boldsymbol{\lambda})}\mathrm{d}\mathbf{r}, \quad \boldsymbol{\lambda} \in \mathrm{R}^2. \tag{6.1}$$

From the fact that the object function f is finitely supported, it follows that its Fourier transform is infinitely supported. Therefore, its inverse Fourier is expressed by

$$f(\mathbf{r}) = \frac{1}{4\pi^2} \int_{\mathrm{R}^2} \hat{f}(\boldsymbol{\lambda})\mathrm{e}^{\mathrm{i}(\mathbf{r}\cdot\boldsymbol{\lambda})}\mathrm{d}\boldsymbol{\lambda}, \tag{6.2}$$

where the integration is carried out over the whole plane R^2. The Radon transform of the function f can be defined by

$$p(a,b) = \int_{\mathrm{D}\cap L(a,b)} f(\mathbf{r})\mathrm{d}\mathbf{r}, \tag{6.3}$$

where $L(a,b)$ is the line described by parameters a, b. The type of these parameters defines the type of the parameterization of the Radon transform. Commonly used is a ϕt-parameterization:

$$L(\phi,t) := \{(x,y) : x\cos\phi + y\sin\phi = t\}. \tag{6.4}$$

If parameter ϕ is fixed, then $L(\phi,t)$ generates a family of parallel lines. According to this, the parallel Radon projection in the direction ϕ is defined as

$$p(\phi,t) = \int_{\mathrm{D}\cap L(\phi,t)} f(\mathbf{r})\mathrm{d}\mathbf{r} = \int_{-\sqrt{R^2-t^2}}^{\sqrt{R^2-t^2}} f(t\tau_\phi + s\tau_\phi^\perp)\mathrm{d}s, \quad -R \le t \le R \tag{6.5}$$

(see Fig. 6.1, left).

Depending on the situation, different notations, the sense of which will be always clear from the context, will be used, e.g., $p_\phi(t)$, $p_\phi(f;t)$, or $p(f;\phi,t)$. If alternatively instead of t the angle parameter ψ is used (see Fig. 6.1, right), we speak of $\phi\psi$ parameterization. The line within this parameterization is expressed by

$$x\cos\phi + y\sin\phi = R\cos\psi. \tag{6.6}$$

Let $x_\alpha y_\alpha$ be coordinates in the coordinate system that is rotated relative to xy at angle α:

$$\begin{aligned} x_\alpha &= x\cos\alpha + y\sin\alpha, \\ y_\alpha &= -x\sin\alpha + y\cos\alpha. \end{aligned} \tag{6.7}$$

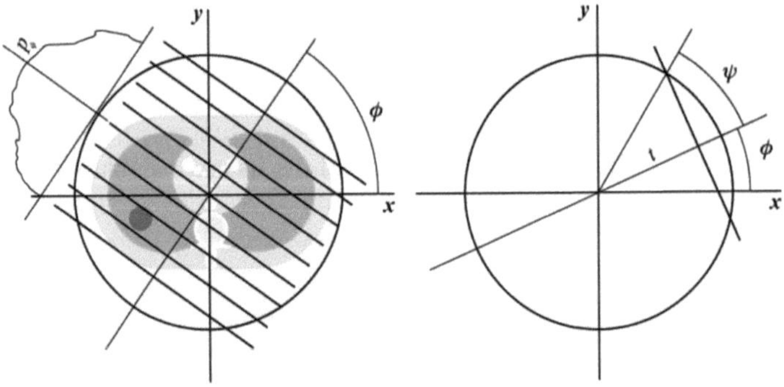

Fig. 6.1 Parallel projection in direction ϕ (*left*); ϕ, t parameterization (*right*)

We speak of $\alpha\psi$ and αY parameterizations of the Radon transform if

$$L(\alpha,\psi) := \{(x,y) : x_\alpha \cos\psi + y_\alpha \sin\psi = R\cos\psi\} \tag{6.8}$$

and

$$L(\alpha,Y) := \{(x,y) : Yx_\alpha + Ry_\alpha = RY\}, \tag{6.9}$$

respectively. Families $L(\alpha,\psi)$ and $L(\alpha,Y)$ where only α is fixed generate fans of rays with an apex lying on the boundary of the disk D. Therefore, both (6.8) and (6.9) are appropriate for the description of data collected over rays emitted by a point source moving along the boundary of the disk (Fig. 6.2).

One of the important problems in practical tomography is the identification of optimal sampling conditions that must be imposed on the Radon transform. Because of limitations of the data acquisition system of the scanner, only discretized approximation of the Radon transform can be obtained. Therefore, on the one side, it is important to know how many data have to be collected in order to fully exhaust the potential of the scanner. On the other side, the radiation dose must be as small as possible, and therefore, the sparsest sampling scheme has to be identified among all of those which ensure the best approximation of the Radon transform. In connection to this problem, the Radon transform is supposed to be sampled on the lattice **L**, the elements of which are discrete points of the plane R^2 coordinated by parameters of the Radon transform (e.g., parameters ϕ, t). So-called admissible lattice can be generated by a 2×2 matrix **W** such that $\mathbf{L} = \mathbf{W}\mathrm{Z}^2$, i.e., for any $i, j \in \mathrm{Z}$,

$$\boldsymbol{\xi} = \mathbf{W}\begin{pmatrix} i \\ j \end{pmatrix}, \quad \boldsymbol{\xi} \in \mathbf{L}. \tag{6.10}$$

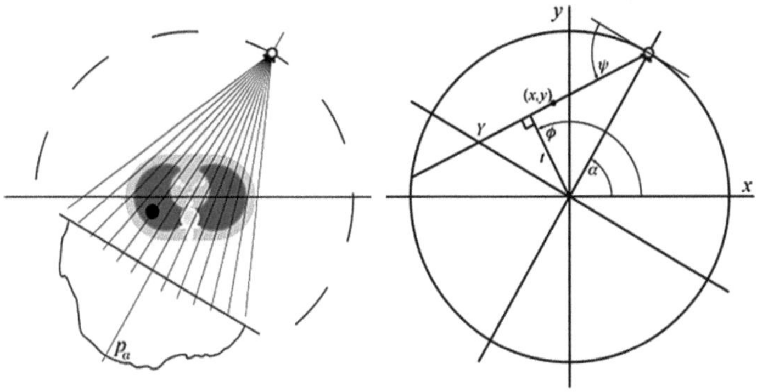

Fig. 6.2 Fan-beam projection (*left*); setup for fan-beam geometry (*right*)

Lattice $\mathbf{L}^{\perp}$ is called dual or reciprocal to the lattice $\mathbf{L}$ if for any $\boldsymbol{\xi} \in \mathbf{L}$ and any $\boldsymbol{\eta} \in \mathbf{L}^{\perp}$, the scalar product $(\boldsymbol{\xi}, \boldsymbol{\eta}) = 2\pi k,\ k \in \mathrm{Z}$. The generator matrix of the reciprocal lattice is

$$\mathbf{W}^{\perp} = 2\pi(\mathbf{W}^{-1})^{T}, \tag{6.11}$$

where the superscript T means transposition (see, e.g., [2]). Let the power spectrum of a function g be supported on the compact region K. The fundamental result of the sampling theory is as follows: if regions K and $K + \boldsymbol{\eta}$ are disjoint for any $\eta \in \mathbf{L}^{\perp}$, then $g(\mathbf{r}) = \sum_{\xi \in \mathrm{L}} g(\boldsymbol{\xi})\tilde{\chi}_K(\mathbf{r} - \boldsymbol{\xi})$, where $\tilde{\chi}_K$ is the inverse Fourier transform of the function $\chi_K, \chi_K(\boldsymbol{\lambda}) = 1$ if $\boldsymbol{\lambda} \in K$, and otherwise $\chi_K(\boldsymbol{\lambda}) = 0$ (see [3]). In other words, the band-limited function can be exactly recovered from its discrete samples if the sampling conditions stated above are satisfied. From the definition of the reciprocal lattice, it follows that the denser $\mathbf{L}^{\perp}$ is, the sparser is $\mathbf{L}$ and vice versa. Therefore, if we want the sampling lattice $\mathbf{L}$ to be as sparse as possible, we have to find such a reciprocal lattice $\mathbf{L}^{\perp}$ that allows us to pack R_2 with mutually disjoint replicas of the K-region as tightly as possible. Certainly, the crucial role in doing this plays the shape of the K-region.

In some parts of the text, concepts from the theory of distributions (see, e.g., [4, 5]) are used. Within the framework of this theory, one defines the set B of test functions. The distribution f over this set is to be understood as a map $f : \mathrm{B} \to \mathrm{R}$ defined as follows: for $h \in \mathrm{B}$,

$$f[h] = \langle f, h \rangle = \int f(x)\overline{h(x)}\mathrm{d}x. \tag{6.12}$$

In the imaging science, such an approach is justified, e.g., when the output of the imaging system can be described in terms of the convolution with the impulse response $\hat{\varphi}$. Let the function $\varphi : \mathrm{R} \to \mathrm{R}$ be smooth, symmetric, and compactly supported, and let its Fourier transform $\hat{\varphi}$ be essentially supported[1] on the interval $[-w, w]$. Beyond this interval, $\hat{\varphi}(\omega)$ is required to decay faster than any degree of $|\omega|^{-1}$. We define the set B as a family $\{\varphi_t, t \in \mathrm{R}\}$ of translates of the impulse response e, i.e., $\varphi_t(s) = \varphi(s-t)$. The set of Fourier transforms of test functions is $\bar{\mathrm{B}} = \{\hat{\varphi}_t, t \in \mathrm{R}\}$, where

$$\hat{\varphi}_t(\omega) = \hat{\varphi}(\omega)\exp(-\mathrm{i}\omega t). \tag{6.13}$$

Then the Fourier transform of the distribution is defined as a map $\hat{f} : \bar{\mathrm{B}} \to \mathrm{R}$ such that $\langle \hat{f}, \hat{h}_t \rangle = 2\pi \langle f, h_t \rangle$ for any $t \in \mathrm{R}$.

6.3 Convolution-Based Algorithms

6.3.1 Parallel-Beam Geometry

The ϕt parameterization is the most clear among all useful parameterizations of the Radon transform mentioned above. Being intensively investigated on the eve of practical tomography, the inversion of the Radon data collected within the parallel-beam geometry is fully understood today. Some important notions, such as, e.g., the resolution of the reconstruction, have been introduced and described for this kind of data geometry. Also the practical fan-beam inversion formulae are the result of the re-parameterization of the inversion formula obtained for the parallel-beam geometry.

Let $p_\phi(t)$ be the Radon projection of the object function f measured at angle ϕ. Due to (6.5), its 1D Fourier transform with respect to t is

$$\hat{p}_\phi(\omega) = \int_{-R}^{R} \int_{-\sqrt{R^2-t^2}}^{\sqrt{R^2-t^2}} f(t\boldsymbol{\tau}_\phi + s\boldsymbol{\tau}_\phi^\perp)\mathrm{e}^{-\mathrm{i}\omega t}\mathrm{d}s\mathrm{d}t. \tag{6.14}$$

Change of variables $\mathbf{r} = t\boldsymbol{\tau}_\phi + s\boldsymbol{\tau}_\phi^\perp$, where $r = (x, y)$, in the double integral of (6.14) yields

$$\hat{p}_\phi(\omega) = \hat{f}(\omega\cos\phi, \omega\sin\phi). \tag{6.15}$$

The identity (6.15) is the statement of the so-called Fourier slice theorem. By taking the inverse Fourier transform, this identity can be rearranged to

$$f(x, y) = \frac{1}{8\pi^2} \int_0^{2\pi} \int_{-\infty}^{\infty} |\omega| \hat{p}_\phi(\omega) e^{\mathrm{i}\omega(\mathbf{r}\cdot\boldsymbol{\tau}_\phi)} \mathrm{d}\omega \mathrm{d}\phi \tag{6.16}$$

[1] That is, $\hat{\varphi}(\omega) \approx 0$ if $\omega \in [-w, w]$.

(see [6] for details). The formula (6.16) makes little sense if the product $|\omega|\hat{p}_\phi(\omega)$ is not absolutely integrable, what can be, e.g., in the case if $\hat{p}_\phi(\omega)\sim\omega^{-2+\varepsilon}$, $\varepsilon \geq 0$. In order to stabilize it, let us multiply both sides of (6.15) with the term $|\omega|$ and handle the resulting identity in terms of distributions over the set B defined above. That is,

$$\left\langle |\omega|\hat{p}_\phi(\omega), \hat{\varphi}_t(\omega)\right\rangle = \left\langle |\omega|\hat{f}(\omega\cos\phi, \omega\sin\phi), \hat{\varphi}_t(\omega)\right\rangle, \quad t \in \mathrm{R} \tag{6.17}$$

or, written in explicit form,

$$\int_{-\infty}^{\infty} |\omega|\hat{p}_\phi(\omega)\hat{\varphi}(\omega)\mathrm{e}^{\mathrm{i}\omega t}\mathrm{d}\omega = \int_{-\infty}^{\infty} |\omega|\,\hat{f}(\omega\cos\phi, \omega\sin\phi)\hat{\varphi}(\omega)\mathrm{e}^{\mathrm{i}\omega t}\mathrm{d}\omega, \quad t \in \mathrm{R}. \tag{6.18}$$

Setting $t = x\cos\phi + y\sin\phi$, integrating over ϕ from 0 to 2π, rearranging the right-hand side appropriately, and changing the variables from polar to rectangular in the double integral on the right-hand side of (6.18) yield

$$\frac{1}{2}\int_0^{2\pi}\int_{-\infty}^{\infty} |\omega|\hat{p}_\phi(\omega)\hat{\varphi}(\omega)\mathrm{e}^{\mathrm{i}\omega(\mathbf{r}\cdot\boldsymbol{\tau}_\phi)}\mathrm{d}\omega\mathrm{d}\phi = \int_{\mathrm{R}^2} \hat{f}(\boldsymbol{\lambda})\hat{\varphi}(|\boldsymbol{\lambda}|)\mathrm{e}^{\mathrm{i}(\mathbf{r}\cdot\boldsymbol{\lambda})}\mathrm{d}\boldsymbol{\lambda}. \tag{6.19}$$

The right integral in (6.19) is the inverse Fourier transform (up to the term $1/4\pi^2$) of the 2D convolution $f * E$, where E is a radial function such that $E(\mathbf{r}) = \varphi(|\mathbf{r}|)$. Hence, we obtain the inversion formula

$$F(x, y) = \frac{1}{8\pi^2}\int_0^{2\pi}\int_{-\infty}^{\infty} |\omega|\hat{P}_\phi(\omega)\mathrm{e}^{\mathrm{i}\omega(\mathbf{r}\cdot\boldsymbol{\tau}_\phi)}\mathrm{d}\omega\mathrm{d}\phi, \tag{6.20}$$

which has the same structure as (6.16), but

$$F(x, y) = f * E(x, y), \tag{6.21}$$

$$P_\phi(t) = p_\phi * \varphi(t). \tag{6.22}$$

Owing to the conditions imposed on φ (see (6.13) and related discussion in Sect. 6.1), the formula (6.20) makes sense for any object function. Clearly, in terms of convolution, the formula (6.20) takes a form given by

$$F(x, y) = \frac{1}{4\pi}\int_0^{2\pi}\int_{-R}^{R} P(\phi, t)H(t' - t)\mathrm{d}t\mathrm{d}\phi, \tag{6.23}$$

where t' is the parameter of the ray that goes through the point (x, y) and

$$H(t) = \frac{1}{2\pi}\int_{-w}^{w} |\omega|\mathrm{e}^{\mathrm{i}\omega t}\mathrm{d}\omega. \tag{6.24}$$

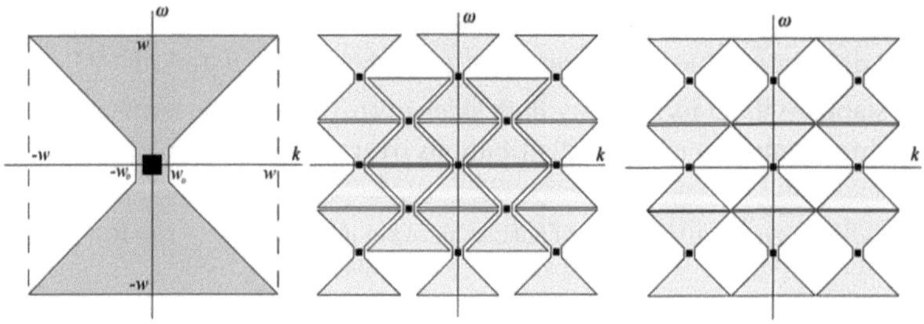

Fig. 6.3 *Left*: K-region of $p(F;\phi,t)$; w is the frequency bound and $w_0 = (1/\vartheta - 1)w$. *Middle*: replicas of bow-tie K-region on the interlaced lattice $\mathbf{L}_i^{\perp}$. *Right*: replicas of bow-tie K-region on the standard lattice $\mathbf{L}_s^{\perp}$

The formula (6.23) is called filtered back projection (FBP) because the evaluation of F can be accomplished in two steps: the convolution of Radon projections with the kernel H (filtering) and evaluating $F(x,y)$ with averages of $\{P_\phi * H(x\cos\phi + y\sin\phi)\}_{0\leq\phi<2\pi}$ (back projection).

Note that (6.22), where on the right there is a convolution of parallel Radon projections with the impulse response of the imaging system, can be thought of as a model of measuring process. It is easy to check that

$$P_\phi(f;t) = p_\phi(F;t), \tag{6.25}$$

that is, the data obtained with the acquisition system of scanner are equivalent to Radon transform of the band-limited function F defined by (6.21). In [7], it was shown that the Fourier transform of the Radon transform of a band-limited function is essentially supported on the region, which has the shape of bow tie shown in Fig. 6.3, left. The rigorous estimate of this region is given in [8], p. 71:

$$K = \{(k,\omega) : |\omega| < w,\ |k| < \max(|\omega|/\vartheta, (1/\vartheta - 1)w)\}, \tag{6.26}$$

where $0 < \vartheta < 1$ is a parameter that is normally close to 1.

Following the sampling theory, let us define the lattice L:

$$\mathbf{L} = \{(\phi_\nu, t_{j\nu}) : \phi_\nu = 2\pi\nu/N, t_{j\nu} = (j + \nu q/N)\Delta;\ \ 0 \leq q,\nu < N,\ \ j \in \mathrm{Z}\}, \tag{6.27}$$

where N is supposed to be even, i.e., $N = 2p$. As it follows from this definition, there are N directions ϕ_ν; for each direction, integrals over an equidistant set of parallel lines with spacing Δ are measured; the collection of parallel lines is shifted by an amount $\nu q/N$ which varies with the angle ϕ_ν. As mentioned above, $\mathbf{L} = \mathbf{W}\mathrm{Z}^2$ where $\mathbf{W}$ is a generator matrix and $\mathrm{Z}^2 = \{(\nu, j),\ \nu, j \in \mathrm{Z}\}$. From (6.27) it follows that

$$\mathbf{W} = \begin{pmatrix} 2\pi/N & 0 \\ q\Delta/N & \Delta \end{pmatrix}. \tag{6.28}$$

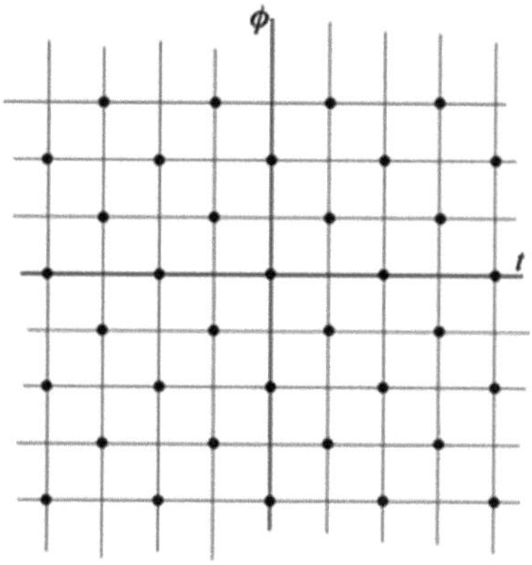

Fig. 6.4 The interlaced lattice (*black circles*) as a subset of the standard lattice (all intersection points). Both lattices have the same theoretical resolution

Then, the generator matrix for the reciprocal lattice is

$$\mathbf{W}^{\perp} = \begin{pmatrix} N & -q \\ 0 & 2\pi/\Delta \end{pmatrix} \tag{6.29}$$

[see (6.11)]. Therefore, for the reciprocal lattice $\mathbf{L}^{\perp}$, we have the following representation:

$$\mathbf{L}^{\perp} = \left\{ \left(Nm - qk, \frac{2\pi k}{\Delta} \right); \quad m, k \in \mathrm{Z} \right\}. \tag{6.30}$$

We will discuss two sampling lattices, namely, the standard lattice $\mathbf{L}_s$ ($q = 0$) and the interlaced lattice $\mathbf{L}_i$ ($q = p$). In the case of the standard sampling, the replicas of the K-region are mutually disjoint if

$$p \geq w/\vartheta, \quad \Delta \leq \pi/w \tag{6.31}$$

(see Fig. 6.3, right). Similarly, sampling conditions for the interlaced lattice are

$$p \geq w(2 - \vartheta)/\vartheta, \quad \Delta \leq 2\pi/w \tag{6.32}$$

(see Fig. 6.3, middle). From (6.31) and (6.32), we see that the interlaced sampling lattice $\mathbf{L}_i$ is almost twice as sparser as the standard lattice $\mathbf{L}_s$. If $\vartheta = 1$, then the interlaced sampling lattice can be obtained from the standard sampling lattice simply by dropping those grid points for which $v + j$ is odd (Fig. 6.4). In other words, the same resolution as that of the data measured on the lattice $\mathbf{L}_s$ can be obtained with only one-half of the data by using the parallel-interlaced scanning geometry. In order to reconstruct data collected over the interlaced lattice, one could first interpolate the missing data to a denser standard lattice and then use (6.23) discretized on this denser lattice. The second approach would be to reconstruct directly from interlaced data (see [9] for error estimates of the corresponding reconstruction).

Practical application of the parallel-beam geometry can be found, e.g., in SPECT (single-photon emission computer tomography). The parallel-beam geometry of data collected with the SPECT device is a result of the collimation that is required for the correct association of the measured data with lines of integration. It has to be noted that because of the poorness of SPECT data and due to attenuation of the emitted photons in the object on their way to the detector, the FBP algorithm happens to be inappropriate. Different algorithms, mainly from the class of iterative methods, have been developed for the reconstruction of SPECT data (see review [10]).

In transmission tomography, the parallel-beam data can be collected with a pencil beam scanner of the first generation. The corresponding scanning process is a combination of N linear translations of a source-detector system across the object, each made along the line of the predefined orientation. The main disadvantage of this process is that data have to be literally built together in a "sample by sample" fashion, which results in a very long acquisition time.

6.3.2 Fan-Beam Geometry

Collecting fan-beam projections (see Fig. 3.2, left) is preferable in computed tomography because of a much shorter acquisition time compared to that required for the collection of parallel-beam projections.

Thorough derivation of corresponding inversion formulae can be found in [11] and [12]. Below, with minor modifications, we reproduce the derivation shown in [6].

The setup corresponding to a fan-beam geometry is as follows: the source moves around the object in the plane xy on the boundary of the disk D. The center of the disk coincides with the origin of the coordinate system. If α is the parameter describing the position of the source, then at $\alpha = 0$ the source is located on the x-axis. The coordinate system $x_\alpha y_\alpha$ defined by (6.7) rotates together with the source.

As it was mentioned above, Radon data collected with the point source can be efficiently described with parameters either αY or $\alpha\psi$ (see (6.8) and (6.9) and Fig. 6.2, right). The fan beam, the rays of which are parameterized with ψ, is called equiangular. If the rays within the fan beam are parameterized with Y, the fan beam is said to be equidistant. In both cases, in order to obtain the inversion formulae, the formula (6.23) has to be re-parameterized properly.

In order to derive the formula for the inversion of the equidistant fan-beam data, the variables (ϕ, t) in the double integral of (6.23) have to be changed for the variables (α, Y) according to the transformation:

$$\phi = \alpha + \cot^{-1}\left(\frac{Y}{R}\right) \tag{6.33}$$

$$t = \frac{RY}{\sqrt{R^2 + Y^2}}. \tag{6.34}$$

Taking into account that

$$t' = \frac{Yx_\alpha + Ry_\alpha}{\sqrt{R^2 + Y^2}} \tag{6.35}$$

and scaling ω, $\omega' = (R - x_\alpha)\omega/\sqrt{R^2 + Y^2}$ yield

$$F(x,y) = \frac{1}{4\pi}\int_0^{2\pi} \frac{R^2}{(R-x_\alpha)^2}\int_{-\infty}^{\infty} \frac{R}{\sqrt{R^2+Y^2}} P_\alpha(Y)\frac{1}{2\pi}\int_{-w}^{w} |\omega|\exp\left(\mathrm{i}\omega\left(\frac{Ry_\alpha}{R-x_\alpha}-Y\right)\right)\mathrm{d}\omega\mathrm{d}Y\mathrm{d}\alpha . \tag{6.36}$$

The parameter Y' of the ray going through the point (x,y) is $Y' = Ry_\alpha/(R - x_\alpha)$ Therefore,

$$F(x,y) = \frac{1}{4\pi}\int_0^{2\pi} \left(\frac{R^2}{R - x_\alpha}\right)^2 \int_{-\infty}^{\infty} \frac{R}{\sqrt{R^2 + Y^2}} P_\alpha(Y)H(Y' - Y)\mathrm{d}Y\mathrm{d}\alpha, \tag{6.37}$$

where H is given by (6.24).

In order to derive the inversion formula for the equiangular fan beam, we have to change variables from (ϕ, t) to (α, ψ) in the double integral of (6.23) according to

$$\phi = \alpha + \psi, \tag{6.38}$$

$$t = R\cos\psi. \tag{6.39}$$

With accounting for the equality $t' = x_\alpha \cos\psi + y_\alpha \sin\psi$, this yields

$$F(x,y) = \frac{1}{8\pi^2}\int_0^{2\pi}\int_0^{\pi} P_\alpha(\psi)\int_{-w}^{w} |\omega|\exp(\mathrm{i}\omega(y_\alpha\sin\psi - (R-x_\alpha)\cos\psi))\mathrm{d}\omega R\sin\psi\mathrm{d}\psi\mathrm{d}\alpha. \tag{6.40}$$

Let $L = L(x,y,\alpha)$ be the distance from the current location of the source to a point (x,y), and let ψ' be the parameter of the ray that goes through this point. Then

$$\begin{aligned} R - x_\alpha &= L\sin\psi', \\ y_\alpha &= L\cos\psi'. \end{aligned} \tag{6.41}$$

After substitution of (6.41) into (6.40) and scaling $\omega' = \omega L$, one obtains

$$F(x,y) = \frac{1}{4\pi}\int_0^{2\pi} \frac{1}{L^2}\int_0^{\pi} P_\alpha(\psi)R\sin\psi\left(\frac{1}{2\pi}\int_{-w}^{w} |\omega|\exp(\mathrm{i}\omega\sin(\psi' - \psi))\mathrm{d}\omega\right)\mathrm{d}\psi\mathrm{d}\alpha \tag{6.42}$$

or equivalently

$$F(x,y) = \frac{1}{4\pi}\int_0^{2\pi}\frac{1}{L^2}\int_0^{\pi} P_\alpha(\psi)R\sin\psi\, g(\psi' - \psi)\mathrm{d}\psi\mathrm{d}\alpha, \tag{6.43}$$

where

$$g(\psi) = H(\sin\psi) \tag{6.44}$$

and H is given by (6.24).

The formula (6.37) can be applied for the inversion of the fan-beam data collected with a flat-panel detector arranged parallel to the y_α-axis. The formula (6.43) is supposed to be used for the reconstruction of the fan-beam data collected with an arc detector. This detector can have the shape of an arc that belongs either to the circle centered in the point source or to the circle the circumference of which contains the point source.

In practice, only discrete samples of the fan-beam data are available. The optimal sampling conditions for the fan-beam scanning geometry are derived in a similar way as in the case of the parallel-beam scanning geometry. A comprehensive study of this topic can be found in [13]. In particular in that work, it was shown that the optimal sampling lattice of the fan-beam Radon transform is equivalent to the sampling lattice of the interlaced parallel geometry with a sinusoidal lateral sampling. The latter is equivalent to the parallel-interlaced lattice on the $\phi\psi$-plane where ψ is defined in (6.6).

6.4 Algorithms Based on Expansion in Basis of Orthogonal Polynomials

In order to obtain the inversion formula within the approach based on the Fourier transform, a special relationship between the object function and its Radon transform in the frequency domain was used [see (6.15)]. Representing a function in the basis of orthogonal polynomials, completely different inversion formulae can be obtained.

Let the object function vanish outside the unit disk D (radius of the disk $R = 1$). Let $\{\varphi_{jk}\}$ be a system of ridge polynomials defined on D, i.e., such that $\varphi_{jk} : \mathrm{D} \to \mathrm{R}$ and

$$\varphi_{jk}(x,y) = \varphi_k(x\cos\theta_j + y\sin\theta_j), \tag{6.45}$$

where $\varphi_k : [-1,1] \to \mathrm{R}$ is a polynomial of degree k. Let φ_{jk} be orthogonal on D with respect to the unit weight function.[2] Finally, let $U_k(t)$ be the Chebyshev polynomial of the second kind:

[2] Polynomial P of degree n is orthogonal on D with respect to the weight function μ if $\int_{\mathrm{D}} P(x,y)Q(x,y)\mu(x,y)\mathrm{d}x\mathrm{d}y = 0$ for any polynomial Q of degree less than n.

$$U_k(t) = \frac{\sin(k+1)\theta}{\sin\theta}, \quad t = \cos\theta. \tag{6.46}$$

The fundamental role within this approach is played by the relation

$$p_\phi(Q;t) = \frac{2}{k+1}\sqrt{1-t^2}U_k(t)Q(\cos\phi, \sin\phi), \tag{6.47}$$

which holds for any orthogonal polynomial Q of degree k (see [14, 15]). Using this relation, one can find that

$$\int_D \varphi_{jk}(x,y)\varphi_{sk}(x,y)\mathrm{d}x\mathrm{d}y = \frac{2}{k+1}\varphi_k(\cos(\theta_j-\theta_s))\int_{-1}^{1} U_k(t)\varphi_k(t)\sqrt{1-t^2}\mathrm{d}t. \tag{6.48}$$

It is known that Chebyshev polynomials are orthonormal on the interval $[-1,1]$ with respect to the weight function $\sqrt{1-t^2}$, i.e.,

$$\int_{-1}^{1} U_m(t)U_n(t)\sqrt{1-t^2}\mathrm{d}t = \frac{\pi}{2}\delta_{m,n}. \tag{6.49}$$

Therefore, inserting $\varphi_k = U_k$ into (6.48) yields

$$\frac{1}{\pi}\int_D U_{jk}(x,y)U_{sk}(x,y)\mathrm{d}x\mathrm{d}y = \langle U_{jk}, U_{sk}\rangle = \frac{1}{k+1}\frac{\sin(k+1)(\theta_j-\theta_s)}{\sin(\theta_j-\theta_s)}, \tag{6.50}$$

where $U_{jk}(x,y) = U_k(x\cos\theta_j + y\sin\theta_j)$. In particular for

$$\theta_j = \theta_{jk} = \frac{j\pi}{k+1}, \tag{6.51}$$

one obtains that $\langle U_{jk}, U_{sk}\rangle = \delta_{s,j}$. Therefore, the set

$$P_n := \{U_{jk}, \quad 0 \le k \le n, \quad 0 \le j \le k\} \tag{6.52}$$

generates an orthonormal basis in the space $\mathbf{\Pi}_n^2$ of bivariate polynomials of degree n. In order to check this, it is sufficient to check that the number of elements in P_n equals $\dim \mathbf{\Pi}_n^2$. This follows from the fact that the coefficients of any polynomial from $\mathbf{\Pi}_n^2$ can be indexed in the same way as in (6.52).

It is known that any function from $\mathrm{L}_2(\mathrm{D})$ can be approximated by polynomials with any degree of accuracy (see, e.g., [16]). This means that P_∞ is complete in $\mathrm{L}_2(\mathrm{D})$, that is, for any $f \in \mathrm{L}_2(\mathrm{D})$, the expansion

$$f(x,y)=\sum_{k=0}^{\infty}\sum_{j=0}^{k}c_k(\theta_{jk})U_k(x\cos\theta_{jk}+y\sin\theta_{jk}), \tag{6.53}$$

$$c_k(\theta_{jk})=\frac{1}{\pi}\int_{\mathrm{D}}f(x,y)U_k(x\cos\theta_{jk}+y\sin\theta_{jk})\mathrm{d}x\mathrm{d}y \tag{6.54}$$

is valid. Truncation of the series (6.53) up to $k=n$ is equivalent to the orthogonal projection onto Π_n^2. Therefore,

$$A_nf(x,y)=\sum_{k=0}^{n}\sum_{j=0}^{k}c_k(\theta_{jk})U_k(x\cos\theta_{jk}+y\sin\theta_{jk}) \tag{6.55}$$

is the best polynomial of degree n approximating the function f in the sense of norm in $\mathrm{L}_2(\mathrm{D})$.[3] Using change of variables $\mathbf{r}=t\boldsymbol{\tau}+s\boldsymbol{\tau}^{\perp}$, where $\boldsymbol{\tau}=(\cos\theta_{jk},\sin\theta_{jk})$, as well as the property expressed by (6.5), we can replace (6.54) by

$$c_k(\theta_{jk})=\frac{1}{\pi}\int_{-1}^{1}p(\theta_{jk},t)U_k(t)\mathrm{d}t. \tag{6.56}$$

Therefore, (6.55) can be used for the inversion of Radon projections measured at specific angles θ_{jk} defined in (6.51). Using the interpolating properties of polynomials, a more general inversion formula can be obtained. Note that the function $g(\phi)=c_k(\phi)U_k(x\cos\phi+y\sin\phi)$ is a trigonometric polynomial of order $2k$. In order to see this, we can apply the operator $p_\phi(\cdot;t)$ to the expansion (6.53), use (6.47), and, as a result, obtain the equality

$$c_k(\phi)=\sum_{j=0}^{k}c_k(\theta_{jk})\frac{1}{k+1}\frac{\sin(k+1)(\phi-\theta_{jk})}{\sin(\phi-\theta_{jk})}, \tag{6.57}$$

where on the right-hand side there is a trigonometric polynomial of order k. Using the fact that $g(\phi)$ is π-periodic and noting that the quadrature formula

$$\int_0^{\pi}g(\phi)\mathrm{d}\phi\approx\frac{\pi}{k+1}\sum_{j=0}^{k}g\left(\frac{j\pi}{k+1}\right) \tag{6.58}$$

is exact for π-periodic trigonometric polynomials of the order not exceeding $2k$, we conclude that

$$\sum_{j=0}^{k}c_k(\theta_{jk})U_k(x\cos\theta_{jk}+y\sin\theta_{jk})=\frac{k+1}{\pi}\int_0^{\pi}c_k(\phi)U_k(x\cos\phi+y\sin\phi)\mathrm{d}\phi. \tag{6.59}$$

[3] That is, $||A_nf-f||_2=\min_{g\in\Pi_n^2}||g-f||_2$.

Substituting this in (6.53) yields

$$f(x,y)=\sum_{k=0}^{\infty}(k+1)\frac{1}{\pi}\int_{0}^{\pi}c_k(\phi)U_k(x\cos\phi+y\sin\phi)\mathrm{d}\phi, \tag{6.60}$$

$$c_k(\phi)=\frac{1}{\pi}\int_{-1}^{1}p(\phi,t)U_k(t)\mathrm{d}t. \tag{6.61}$$

Following simple considerations, we come up to conclusion according to which the best approximation of Radon projection p_ϕ, which can be obtained with detector of limited resolution, is given by trigonometric polynomial.[4] In the following, the degree of this polynomial will be referred to as resolution of the detector. Hence, if projections are collected with the detector of the resolution n, we can set coefficients $c_k(\phi)$ to zero for all $k\geq n$. As a consequence, the series (6.60) can be truncated to k. Let us introduce angles ϕ_v:

$$\phi_v=\frac{\pi v}{n},\quad 0\leq v<n. \tag{6.62}$$

Noting that the Gaussian quadrature

$$\frac{1}{\pi}\int_{0}^{\pi}c_k(\phi)U_k(x\cos\phi+y\sin\phi)\mathrm{d}\phi\approx\frac{1}{n}\sum_{v=0}^{n-1}c_k(\phi_v)U_k(x\cos\phi_v+y\sin\phi_v) \tag{6.63}$$

is exact for $k<n$, after truncation of (6.60) up to $k<n$, one obtains the formula

$$f(x,y)\approx A_{n-1}f(x,y)=\frac{1}{n}\sum_{k=0}^{n-1}\sum_{v=0}^{n-1}(k+1)c_k(\phi_v)U_k(x\cos\phi_v+y\sin\phi_v), \tag{6.64}$$

which can be used for recovering the polynomial A_{n-1} introduced in (6.55) from n projections collected with an acquisition system of the resolution n. Due to (6.59), increasing the number of projections further has no effect.

One could also pose the reverse question, namely, what has to be maximum resolution of detector if only n projections are available. From the mathematical point of view, if we want the quadrature (6.63) to be exact, the condition k has to be satisfied. In this case, arguing in the same way as before will end up in the conclusion that the detector has to be n. However, if the resolution of the detector is higher than the number of available projections, we may truncate the series (6.64) to some $m>n$ allowing (6.63) to hold only approximately. Increasing the truncation level of the series (6.53), we increase the order of approximating polynomials.

[4] In this case, the output of acquisition system is the convolution with the impulse response of the ideal filter.

This can improve visual contrast of the reconstructed image, at least, to a certain degree that depends on the acceptable degree of approximation of (6.63).

The inversion formula (6.64) is called OPED (orthogonal polynomial expansion on disk) due to the formalism that has been used for its derivation. As one can see from the structure of OPED, the Radon data appear in form of projections measured at discrete angles (6.62). In practice, projections are not continuous either. Therefore, in order to calculate the coefficients $c_k(\phi_v)$, the integral

$$\int_{-1}^{1} p_\phi(t) U_k(t) \mathrm{d}t = \int_{0}^{\pi} p_\phi(\psi) \sin(k+1)\psi \mathrm{d}\psi \tag{6.65}$$

has to be approximated by an appropriate quadrature. The best result is expected when the quadrature relates to the regular sampling of the parameter ψ (see [17, 18]). If the p_ϕ is a polynomial, then the quadrature is exact. Hence, sampling conditions that must be imposed on the Radon transform are defined by the order of a bivariate polynomial that is supposed to play the role of interpolating polynomial for the object function. From above considerations, it follows that the Radon transform is sampled on lattice $\{\phi_v, \psi_j\}$, where ϕ_v is defined in (6.62) and $\psi_j = \psi_0 + j\,\pi/n$, $j = 0, \ldots, n-1$. Parameter ψ_0 is free to choose. Normally, this parameter is either zero or $\pi/2n$. The corresponding sampling lattice is the standard lattice in the plane coordinated by (ϕ, ψ). In fact, we can set up even sparser sampling condition without compromising the quality of the final result. Due to [14] and [19], the optimal sampling of the Radon transform of a polynomial supported on the unit disk D relates to circle geometry. This scanning geometry is defined by the set of $n(n-1)/2$ linear segments $[X_i, X_j]$, where n distinct points X_i are uniformly distributed on the boundary of the disk. In [14] and [19], it was shown that for $n = 2p$, the set of integrals measured within this scanning geometry determines uniquely the polynomial of degree up to $n-2$. The scanning geometry defined thus coincides with the interlaced lattice on the $\phi\psi$-plane. It is worth to note that the data collected within this scanning geometry can be reordered to fan-beam data and vice versa. Thus, in some cases, OPED can also be used for the reconstruction from data collected within the fan-beam geometry. Clearly, in order to do this, the fan-beam data have to be appropriately reordered.

In practice, the geometry of data collected by positron emission tomography (PET) is apparently a circle geometry (see Fig. 6.5). In the figure, an idealized view of a PET acquisition system is shown.

The set of circularly arranged detectors is divided into two groups of alternating elements. According to this, two groups of data can be defined. The data of the first groups can be associated with lines of response between detectors of the same color, and the data of the second group with lines of response between detectors of different colors. It is not difficult to see that both groups correspond to the standard lattices

$$\mathbf{L}_1 = \{(\phi_v, \psi_j) : \phi_v = \pi v/p, \quad \psi_j = \pi j/p; \quad 0 \le v, j \le p\} \tag{6.66}$$

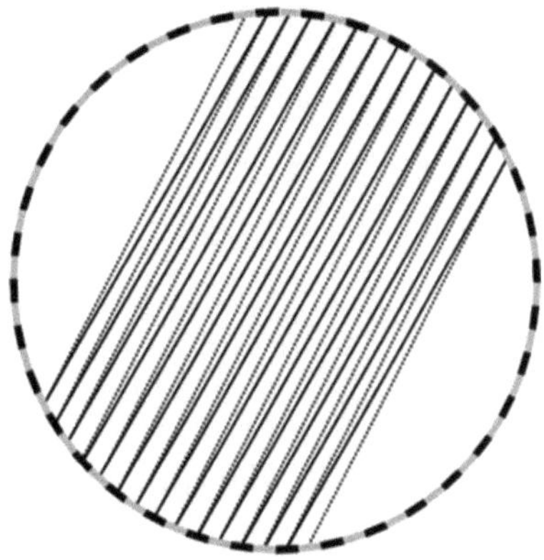

Fig. 6.5 PET detector ring with $2n$ detectors and two parallel sets of lines of response

for the data of the first group and

$$\mathbf{L}_2 = \{(\phi_v, \psi_j) : \phi_v = \pi(v + 1/2)/p, \ \ \psi_j = \pi(j + 1/2)/p; \quad 0 \leq v, \ j \leq p\} \tag{6.67}$$

for the data of the second group. The lattice $\mathbf{L}_1 + \mathbf{L}_2$, that is, the lattice consisting of the grid points both of $\mathbf{L}_1$ and of $\mathbf{L}_2$, generates the interlaced lattice in (ϕ, ψ)-plane.

The same data geometry as the one described above can be realized in transmission tomography with the help of a mask in form of the ring shown in Fig. 6.5, where the detectors are replaced by alternating shields and apertures [20]. The object is supposed to be inside the ring mask, and the source being outside the mask rotates about it. The radiation achieves the object through apertures of the mask. The data can be collected either with the detector array rotating together with the source either inside the mask or outside the mask.

It is also possible to collect such data with help of a scanning system depicted in Fig. 6.6. This system is similar to a source-detector system rotating as a whole around the object. If we allow the detector array to perform an additional rotation about the source but in opposite direction, at a speed that provides fixed spatial orientation of the line connecting a detector element with the source, then the data collected with this detector element would generate a parallel projection at a given projection angle, namely, at the angle that is determined by the orientation of the line connecting the source and the detector element considered [20]. Hence, if detectors are distributed on the arc of the size π around the source, the collection of the set of data required for the reconstruction is completed after the point source has covered the arc of the size $\pi + \beta$ along its trajectory, where β is the angle of the fan beam (Fig. 6.6). In terms of the amount of radiation transmitted through the object, such a scan is equivalent to the short scan of the scanner of third generation. However, the resolution of the data collected within the scanning geometry realized by the system depicted in Fig. 6.6 is higher due to the fact that the obtained data are not redundant.

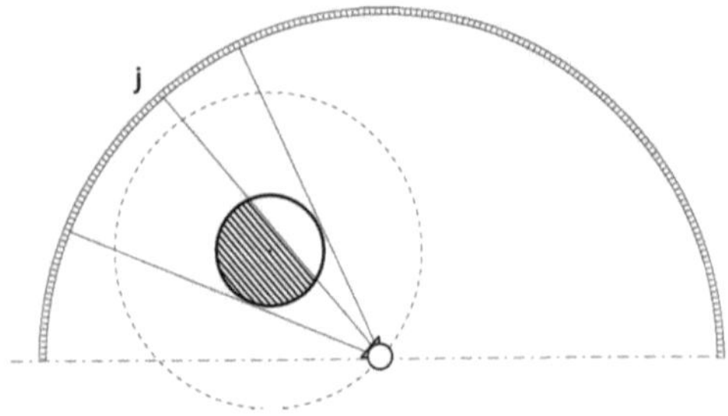

Fig. 6.6 Scanning system allowing the collection of data upon the lattice in the $\phi\psi$-plane

6.5 Back Projection of Derivative of Projections

Another class of algorithms based on differentiated back projection (DBP) takes advantage of the back projection of the derivative of the projections (see [21] and references there). As will be shown below, the result of this step relates to the object through the Hilbert transform, which can be inverted using existing formulae.

The Hilbert transform of a 1D function f is the convolution with the kernel $1/\pi s$:

$$Hf(s) = \int_{-\infty}^{\infty} \frac{f(t)\mathrm{d}t}{\pi(s-t)}. \tag{6.68}$$

It is known (see [22]) that if a 1D function f satisfies the invertibility conditions, according to which its support is strictly inside the interval $(-R, R)$, and the Hilbert transform $Hf(s)$ is known for $s \in [-R, R]$, then the following inversion formula

$$f(s) = \frac{-1}{\sqrt{R^2 - s^2}} \left(\int_{-R}^{R} \frac{Hf(t)\sqrt{R^2 - t^2}}{\pi(s-t)} \mathrm{d}t + C \right) \tag{6.69}$$

can be used for recovering $f(s)$, $s \in (-R, R)$, from Hf, where C is an unknown constant which can be determined from knowledge of $f(s)$ at selected values of s.

In two dimensions the Hilbert transform in a specific direction can be defined. The function $H_\theta f(\mathbf{r})$, $\quad \mathbf{r} \in \mathrm{R}^2$,

$$H_\theta f(\mathbf{r}) = \int_{-\infty}^{\infty} \frac{f_\theta(t,s)\mathrm{d}s}{\pi(s'-s)}, \tag{6.70}$$

where $f_\theta(t,s) = f(t\theta + s\theta^\perp)$, $\boldsymbol{\theta} = (\cos\theta, \sin\theta)$, is called Hilbert transform of the 2D function in direction θ. If for some fixed t the 1D function $f_\theta(t,\cdot)$ satisfies the invertibility conditions formulated above, it can be recovered from $H_\theta f$ using (6.69). It can be shown that

$$-2\pi H_\theta f(\mathbf{r}) = b_\theta(x,y) = \int_{\theta}^{\pi+\theta} p'_\phi(x\cos\phi + y\sin\phi)\mathrm{d}\phi, \tag{6.71}$$

where the prime means derivative. Indeed, representing the derivative of the projection p_ϕ by its Fourier transform and using the Fourier slice theorem, we can obtain the equality

$$b_\theta(x,y) = \frac{1}{2\pi}\int_0^\pi\int_{-\infty}^{\infty} \mathrm{i}\omega\hat{f}(\omega\cos(\phi+\theta),\omega\sin(\phi+\theta))\mathrm{e}^{\mathrm{i}\omega(x\cos(\phi+\theta)+y\sin(\phi+\theta))}\mathrm{d}\omega\mathrm{d}\phi. \tag{6.72}$$

It is not difficult to transform (6.72) to

$$b_\theta(x,y) = \frac{1}{2\pi}\int_0^{2\pi}\int_0^{\infty} i\mathrm{sgn}(\omega\sin\phi)\hat{f}(\omega\cos(\phi+\theta),\omega\sin(\phi+\theta))\mathrm{e}^{\mathrm{i}\omega(x\cos(\phi+\theta)+y\sin(\phi+\theta))}\omega\mathrm{d}\omega\mathrm{d}\phi \tag{6.73}$$

Finally, changing the variables from polar to rectangular and using the identity

$$\hat{f}_\theta(\omega_1,\omega_2) = \hat{f}(\omega_1\boldsymbol{\theta} + \omega_2\boldsymbol{\theta}^\perp) \tag{6.74}$$

yield

$$b_\theta(x,y) = \frac{1}{2\pi}\int_{-\infty}^{\infty}\int_{-\infty}^{\infty} i\mathrm{sgn}(\lambda_2)\,\hat{f}_\theta(\lambda_1,\lambda_2)\mathrm{e}^{\mathrm{i}(\lambda_1 t+\lambda_2 s)}\mathrm{d}\lambda_1\mathrm{d}\lambda_2, \tag{6.75}$$

where $t = \mathbf{r}\cdot\boldsymbol{\theta}, s = \mathbf{r}\cdot\boldsymbol{\theta}^\perp$. In the sense of distributions, the inverse Fourier transform of $i\mathrm{sgn}(\omega)$ is $-1/\pi t$ (see [5], Sect. 2, Table 1). Recalling that the Fourier transform of the convolution of two functions is equal to the product of their Fourier transforms, one obtains (6.71).

Hence, if for each t the function $H_\theta f(t,s)$ satisfies the invertibility conditions with respect to s, the formula (6.69) can be used to recover the function f.

The advantage of this method is that differentiation is essentially a local operation. This is true neither for FBP nor for OPED where the projections have to be transformed with a kernel of unlimited support that cannot be truncated. Therefore, the whole projection must necessarily contribute to the reconstructed point. Under certain conditions, local properties of the DBP can be efficiently used for the reconstruction of a region of interest from projections even if some of them are truncated. To make this clear, let us define an *admissible segment* $T(\theta,t) := \{t\boldsymbol{\theta} + s\boldsymbol{\theta}^\perp, -R \le s \le R\}$ as one that satisfies the following conditions (1) $H_\theta f(\mathbf{r})$ is known for any $\mathbf{r}\in T(\theta,t)$ and (2) $T(\theta,t)$ crosses boundaries of the support D of the object function f. These conditions mean that any point of the segment $T(\theta,t)$ contributes to all projections and that $f_\theta(\mathbf{r}) = 0$ for all $\mathbf{r}\notin T\cap D$. With other words, the invertibility conditions are satisfied on $T(\theta,t)$, and therefore,

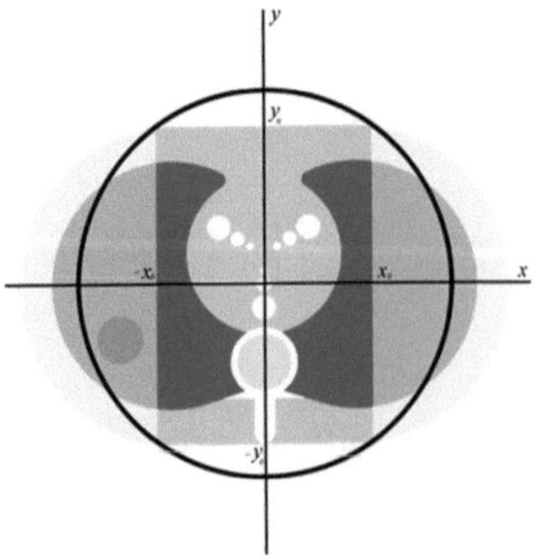

Fig. 6.7 *Black circle* is the field of view. Distinct rectangular part within the *black circle* is the ROI within the object that can be reconstructed. There are no admissible segments for the points inside of fade parts of the object

the 1D function $f_\theta(t,\cdot)$ can be recovered with the help of (6.69). If there are no truncated projections, i.e., the whole object is inside the field of view, then any segment crossing the object is admissible. Otherwise, the set of all admissible segments generates the region inside the object that can be reconstructed. An example is given in Fig. 6.7. In this figure, the reconstructable ROI is the distinct rectangular region $[-x_0, x_0] \times [-y_0, y_0]$ inside the field of view. The segment $T(0,t)$ is admissible if $t \in [-x_0, x_0]$. One can see that there are no admissible segments intersecting the fade region of the object. Indeed, any line intersecting the fade region also intersects that part of the object that lies outside the field of view and hence contains points contributing not to all projections.

Using the setup introduced for the fan-beam geometry in Fig. 6.2, right, it is possible to rewrite the formula (6.71) in fan-beam coordinates. We will do it here for the equiangular fan. The derivation for the equidistant fan can be found in [21].

Due to properties of the Fourier transform, the derivative of the function f can be represented by

$$\frac{\mathrm{d}f(t)}{\mathrm{d}t} = \int_{-\infty}^{\infty} f(s)H(t-s)\mathrm{d}s, \tag{6.76}$$

where $H(t) = \frac{1}{2\pi}\int_{-\infty}^{\infty} \mathrm{i}\omega \exp(\mathrm{i}\omega t)\mathrm{d}\omega$. Therefore, noting that

$$\int_{\theta}^{\pi+\theta} p'(\phi, x\cos\phi + y\sin\phi)\mathrm{d}\phi = \int_0^{2\pi} \operatorname{sgn}(\sin(\phi-\theta))p'(\phi, x\cos\phi + y\sin\phi)\mathrm{d}\phi, \tag{6.77}$$

one obtains the expression

$$b_\theta(x,y) = \int_0^{2\pi} \operatorname{sgn}(\sin(\phi-\theta)) \int_{-R}^{R} p(\phi,t)H(t'-t)\mathrm{d}t\mathrm{d}\phi, \tag{6.78}$$

where $t' = x\cos\phi + y\sin\phi$ and R is the radius of the source trajectory. Changing variables from parallel beam to equiangular fan beam [see (6.39) and (6.40)] yields

$$b_\theta(x,y) = \int_0^{2\pi}\int_0^{\pi} \operatorname{sgn}(\sin(\alpha+\psi-\theta))p(\alpha,\psi)H(L\sin(\psi-\psi'))R\sin\psi \mathrm{d}\psi\mathrm{d}\alpha, \tag{6.79}$$

where ψ' is the parameter that satisfies (6.6). Scaling $\omega = \frac{\psi}{L\sin\psi}\omega'$, we find that $H(L\sin\psi) = (\psi/L\sin\psi)^2 H(\psi)$. Substituting this into (6.79) and taking into account the property (6.76), we obtain

$$b_\theta(x,y) = \int_0^{2\pi} \frac{1}{L^2}\frac{\partial}{\partial\psi}\left(\operatorname{sgn}(\sin(\alpha+\psi-\theta))p(\alpha,\psi)R\sin\psi\left(\frac{\psi-\psi'}{\sin(\psi-\psi')}\right)^2\right)_{\psi=\psi'} \mathrm{d}\alpha. \tag{6.80}$$

It is clear that $(\partial/\partial\psi((\psi'-\psi)/\sin(\psi'-\psi))^2)_{\psi'=\psi} = 0$. Therefore, (6.80) can be reduced to

$$b_\theta(x,y) = \int_0^{2\pi} \frac{1}{L^2}\frac{\partial}{\partial\psi}(\operatorname{sgn}(\sin(\alpha+\psi-\theta))p(\alpha,\psi)R\sin\psi)_{\psi=\psi'}\mathrm{d}\alpha, \tag{6.81}$$

which in turn can be represented by the sum

$$\begin{aligned} b_\theta(x,y) &= \int_0^{2\pi} \frac{1}{L^2}\operatorname{sgn}(\sin(\alpha+\psi'-\theta))\frac{\partial}{\partial\psi}(p(\alpha,\psi)R\sin\psi)_{\psi=\psi'}\mathrm{d}\alpha \\ &+ \int_0^{2\pi} \frac{1}{L^2}p(\alpha,\psi')R\sin\psi'\frac{\partial}{\partial\psi}(\operatorname{sgn}(\sin(\alpha+\psi-\theta)))_{\psi=\psi'}\mathrm{d}\alpha = I_1 + I_2. \end{aligned} \tag{6.82}$$

Let us look at the second term of this sum. The function $\operatorname{sgn}(\sin(\alpha+\psi-\theta))$ has discontinuities for those (α,ψ) that satisfy

$$\alpha+\psi-\theta = \pi l, \quad l \in \mathrm{Z}. \tag{6.83}$$

For the sake of convenience, parameters satisfying (6.83) will be referred to as *singular*, and the related ray will be referred to as *singular ray*. If α_1, ψ_1 are singular, then so are parameters $\alpha_1 + 2\psi_1, \pi - \psi_1$. Obviously, corresponding singular rays lie on the same line, which will be referred to as *singular* as well. It can be shown that there is only one singular line that goes through the point (x,y). Indeed, let

$\mathbf{v} = (x - R\cos\alpha, y - R\sin\alpha)$ be the vector directed from position α of the source to the point (x, y). Unifying (6.83) and (6.6), it can be checked that if α is singular, then $(x - R\cos\alpha)\cos\theta + (y - R\sin\alpha)\sin\theta = 0$. This means that the singular line is perpendicular to the fixed vector $\boldsymbol{\theta} = (\cos\theta, \sin\theta)$. Clearly, there is only one line that goes through (x, y) and is perpendicular to $\boldsymbol{\theta}$. Hence, there are only two singular positions of the source, namely, those where the singular line intersects the source trajectory. Therefore, for the second integral in (6.82), we obtain

$$I_2 = 2\left(\frac{1}{L_1^2} - \frac{1}{L_2^2}\right)p(\alpha_1, \psi_1)R\sin\psi_1, \tag{6.84}$$

where L_1 and L_2 are the distances from the first and the second singular positions of the source to the point (x, y), respectively. Hence,

$$\begin{aligned} H_\theta(t,s) &= \frac{1}{\pi}\left(\frac{1}{L_1^2} - \frac{1}{L_2^2}\right)p(\alpha_1, \psi_1)R\sin\psi_1 \\ &+ \frac{1}{2\pi}\int_0^{2\pi}\frac{1}{L^2}\operatorname{sgn}(\sin(\alpha + \psi' - \theta))\frac{\partial}{\partial\psi}(p(\alpha,\psi)R\sin\psi)_{\psi=\psi'}\mathrm{d}\alpha, \end{aligned} \tag{6.85}$$

where the parameter ψ' satisfies (6.6); $\psi_1 = \pi - \gamma$, $\alpha_1 = \theta + \psi_1$, and γ is such that $R\cos\gamma = x\cos\theta + y\sin\theta$.

Apparently, all facts mentioned above concerning the reconstruction from truncated parallel projections are also valid for truncated fan-beam projections. The difference is only in the way of generating the image of the Hilbert transform $H_\theta(t, s)$.

On the other hand, it is known that fan-beam data collected over an arc of size $l < \pi + \gamma$, where γ is the fan angle, are incomplete in the sense that parameters (ϕ, t) of the rays associated with collected data do not entirely cover the range of parameters. Using the DBP formula, it is possible to reconstruct a region of interest even if the set of fan-beam data is smaller than the minimal set. We demonstrate this with an example shown in Fig. 6.8.

In Fig. 6.8 (left), there is a field of view that corresponds to the fan angle γ. It is supposed that the source trajectory is the bold boundary of the upper semicircle; the right side of the figure shows the diagram built for parallel-beam parameters (ϕ, t). The line $l(x, y) := \{(\phi, t) : x\cos\phi + y\sin\phi = t\}$ is shown in the diagram for the point (x, y) of the field of view. It is clear that points of the upper half of the field of view contribute to all parallel projections. Therefore, any segment that transfers the object in the upper semicircle parallel to the x-axis is admissible. Hence, the Hilbert image $H_{\pi/2}(y, x), x \in \left(-\varepsilon - \sqrt{r^2 - y^2}, \varepsilon + \sqrt{r^2 - y^2}\right), \varepsilon > 0$ can be generated via back projection of derivatives of projections, and then this region can be reconstructed on each distinct admissible segment that is parallel to the x-axis.

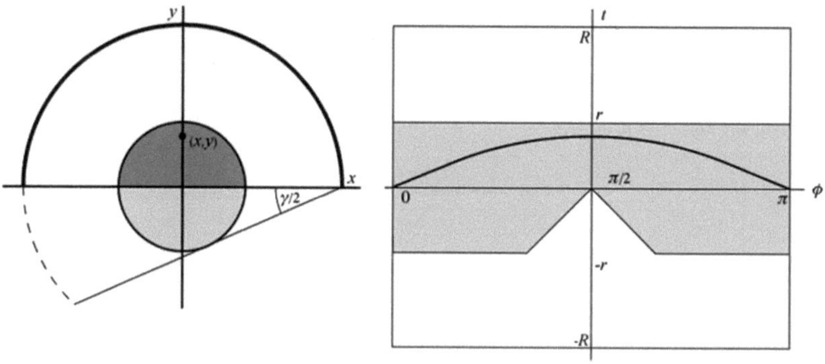

Fig. 6.8 *Left*: trajectory of the point source (*bold semicircle*) and the field of view related to the fan angle γ. *Right*: the (ϕ, t) diagram of data corresponding to the setup depicted on the *left*. The *black curve* corresponds to the point (x, y)

6.6 Extension to 3D

In 3D transmission tomography, one has to deal with cone-beam data that are collected along the source trajectories of a predefined configuration. Especially two configurations, circular and spiral, are of great relevance in clinical practice because they allow operating at a high rotating speed due to their symmetry.

Unfortunately, a circular trajectory does not satisfy Tuy's data sufficiency condition [23] which requires that every plane intersecting the object must also intersect the trajectory of the source at least once. Therefore, only approximate reconstruction from cone-beam projections collected within this geometry is possible. For the reconstruction from these data, the FDK algorithm [24] was introduced. Despite its approximate character, the FDK algorithm is still widely used, either in its original form or in various modifications developed for improved accuracy. Below, the main steps of the derivation of this algorithm are outlined with minor modifications.

The formal setup is very similar to that one introduced above for the derivation of the fan-beam reconstruction formula. The source rotates about the object in the midplane. The coordinate system xyz is fixed; z is the axis of rotation, the xy-plane coincides with the midplane, and the x-axis is directed towards the initial position of the source. That is, for $\alpha = 0$, the source is located on the x-axis. The coordinate system that rotates together with the source is $x_\alpha y_\alpha$ (see (6.7) for the transformation between xy and $x_\alpha y_\alpha$). The detector plane coincides with the plane $y_\alpha z$ but is supplied with its own rectangular coordinates YZ.

Let us rewrite the formula obtained for the inversion of equidistant fan-beam data (6.37) in the form

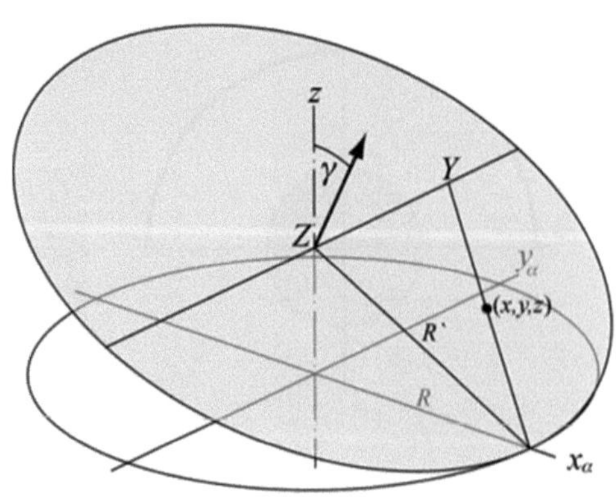

Fig. 6.9 Tilted arrangement. The tilted *gray plane* containing the source and the point (x, y, z) intersects the $y_\alpha z$ -plane along the line $z = Z$

$$f(\mathbf{r})_{z=0} = \frac{1}{4\pi}\int_0^{2\pi} \delta f(\mathbf{r},\alpha)_{z=0},$$

$$\begin{aligned}\delta f(\mathbf{r},\alpha)_{z=0} &= \frac{R^2}{(R-x_\alpha)^2}\int_{-\infty}^{\infty}\frac{R}{\sqrt{R^2+Y^2}}P_\alpha(Y,Z)_{Z=0}\frac{1}{2\pi}\\ &\times\int_{-W}^{W}|\omega|\exp\left(\mathrm{i}\omega\left(\frac{Ry_\alpha}{R-x_\alpha}-Y\right)\right)\mathrm{d}\omega\mathrm{d}Y\mathrm{d}\alpha, \qquad (6.86)\end{aligned}$$

where R is the radius of the source trajectory and $\mathbf{r} = (x, y, z)$. Here it has been specifically noted that $z = 0$ is the plane under consideration. The value $\delta f(\mathbf{r},\alpha)_{z=0}$ is the contribution to the value $f(\mathbf{r})_{z=0}$ made by the projection measured at position α.

The main assumption made while deriving the FDK formula is that if $z \neq 0$ then the value $\delta f(\mathbf{r},\alpha)$ contributes to $f(\mathbf{r})$ in another, tilted arrangement (see Fig. 6.9), where the source instantaneously rotates in the plane that contains the point $\mathbf{r}$.[5] Hence, in order to derive the FDK formula, the expression for $\delta f(\mathbf{r},\alpha)$ has to be obtained for arbitrary z.

Let the source be in position α, and let $\mathbf{r}$ be the point that has to be reconstructed. Due to the main assumption, the source instantaneously moves in the plane that contains $\mathbf{r}$ and intersects the plane $y_\alpha z$ through the line $z = Z$ (see Fig. 6.9). Let R' be the distance from the source to the point $(0, 0, Z)$. This value can be thought of as the radius of the virtual trajectory in the tilted plane. It is clear that

$$R' = \sqrt{R^2 + Z^2}. \qquad (6.87)$$

[5] Clearly, if $z \neq 0$, then for different α, there are different planes that contain this point, so that all planes of projections that contribute to the reconstruction at given $\mathbf{r}$ may be visualized as forming a bundle.

Both trajectories—the real one and the virtual one—are tangent to each other in the source point. Therefore, relation $R\mathrm{d}\alpha = R'\mathrm{d}\alpha'$ holds, which yields

$$\mathrm{d}\alpha' = \frac{R}{R'}\mathrm{d}\alpha. \tag{6.88}$$

For a tilting angle γ,we have

$$\frac{R}{R'} = \cos\gamma. \tag{6.89}$$

Due to the main assumption, all characteristic values participating in (6.86) have to be replaced by their equivalents in the tilted arrangement. In addition to the radius of the trajectory, another characteristic value is the distance d from the source to the line that contains $\mathbf{r}$ and is parallel to y_α. If $z = 0$, i.e., $\mathbf{r}$ is in the midplane, then $d = R - x_\alpha$. Otherwise, $d' = (R - x_\alpha)/\cos\gamma$ or with accounting for (6.89):

$$d' = \frac{R'}{R}(R - x_\alpha). \tag{6.90}$$

Noting that the parameter Y is not changed in the tilted arrangement and taking into account (6.87)–(6.90), one obtains

$$\begin{aligned}\delta f(\mathbf{r},\alpha) = {} & \frac{R^2}{(R-x_\alpha)^2}\int_{-\infty}^{\infty}\frac{R}{\sqrt{R^2+Z^2+Y^2}}P_\alpha(Y,Z)\frac{1}{2\pi} \\ & \times \int_{-w}^{w}|\omega|\exp\left(\mathrm{i}\omega\left(\frac{Ry_\alpha}{R-x_\alpha}-Y\right)\right)\mathrm{d}\omega\mathrm{d}Y\mathrm{d}\alpha.\end{aligned} \tag{6.91}$$

Therefore,

$$f(x,y,z) \approx \frac{1}{4\pi}\int_0^{2\pi}\frac{R^2}{(R-x_\alpha)^2}\int_{-\infty}^{\infty}\frac{R}{\sqrt{R^2+Z^2+Y^2}}P_\alpha(Y,Z)H(Y'-Y)\mathrm{d}Y\mathrm{d}\alpha, \tag{6.92}$$

where H is given by (6.24) and

$$Y' = \frac{Ry_\alpha}{R-x_\alpha} \tag{6.93}$$

is the Y-coordinate of the point where the ray intersects the "detector" plane $y_\alpha z$.

Finally, limitations of the acquisition system in the z-direction have to be taken into account. Note that for any function $F(s)$ with the Nyquist limit w_z, the representation

$$F(s) = \frac{1}{2\pi}\int_{-\infty}^{\infty}\hat{F}(\omega)1_{[-w_z,w_z]}(\omega)\exp(\mathrm{i}\omega s)\mathrm{d}\omega = \int_{-\infty}^{\infty}F(t)\frac{\sin w_z(s-t)}{\pi(s-t)}\mathrm{d}t \tag{6.94}$$

is valid. Let $F(Z)$ be the function appearing in the inner integral of (6.92), i.e.,

$$F(Z) = F_Y(Z) = \frac{R}{\sqrt{R^2 + Z^2 + Y^2}} P_\alpha(Y, Z). \tag{6.95}$$

Representing $F(Z)$ as in (6.94) and substituting it in (6.92) yield

$$f(\mathbf{r}) \approx \frac{1}{4\pi} \int_0^{2\pi} \frac{R^2}{(R - x_\alpha)^2} \int_{-\infty}^{\infty} \int_{-\infty}^{\infty} \frac{R}{\sqrt{R^2 + Z^2 + Y^2}} P_\alpha(Y, Z) g(Z' - Z) H(Y' - Y) \mathrm{d}Z \mathrm{d}Y \mathrm{d}\alpha, \tag{6.96}$$

where

$$g(s) = \frac{\sin w_z s}{\pi s}, \tag{6.97}$$

and (Y', Z'), where $Z' = zR/(R - x_\alpha)$, are the coordinates of the spot on the plane $y_\alpha z$ that is hit by the ray going through the point $\mathbf{r}$.

As stated in [24], the FDK formula (6.96) is exact only if the object function $f(\mathbf{r})$ is independent of z. In general, the FDK reconstruction is exact only in the midplane but exhibits artifacts in slices lying far away from the midplane. The degree of inaccuracy of the FDK reconstruction depends not only on the distance from the midplane but is also highly object dependent. Nevertheless, the reconstruction is sufficiently good if the object occupies a relatively narrow area containing the mid-slice. This circumstance made it possible to successfully adopt the FDK approach for the reconstruction from the cone-beam data collected along the helical trajectory (spiral CT). For this aim numerous extensions and modifications of the FDK algorithms have been proposed in the literature (see, e.g., [25, 26]).

References

1. Matej S, Lewitt RM (1996) Practical consideration for 3-D image reconstruction using spherically symmetric volume elements. IEEE Trans Med Imaging 15:68–78
2. Higgins JR (1996) Sampling theory in Fourier and signal analysis: foundations. Oxford University, New York, NY
3. Petersen DP, Middleton D (1962) Sampling and reconstruction of wave-number limited functions in N-dimensional Euclidean spaces. Inf Control 5:279–323
4. Schwartz L (1950–1951) Théorie des distributions, 1–2. Hermann
5. Gelfand IM, Shilov GE (1964) Generalized functions, vol I. Academic, New York, NY
6. Kak A, Slaney M (2001) Principles of computerized tomographic imaging. SIAM, Philadelphia, PA
7. Rattey PA, Lindgren AG (1981) Sampling the 2-D Radon transform. IEEE Trans Acoust Speech Signal Process ASSP-29:994–1002
8. Natterer F (1986) The mathematics of computerized tomography. Wiley, Chichester

9. Faridani A, Ritman EL (2000) High-resolution computed tomography from efficient sampling. Inverse Probl 16:635–650
10. Lewitt RM, Matej S (2003) Overview of methods for image reconstruction from projections in emission computed tomography. Proc IEEE 91:1588–1611
11. Herman GT, Lakshminarayanan AV, Naparstek A (1976) Convolution reconstruction techniques for divergent beams. Comput Biol Med 6:259–274
12. Lakshminarayanan AV (1975) Reconstruction from divergent ray data. Department of Computer Science technical report TR-92, State University of New York, Buffalo, NY
13. Natterer F (1993) Sampling in fan beam tomography. SIAM J Appl Math 53(2):358–380
14. Marr RB (1974) On the reconstruction of a function on a circular domain from a sampling of its line integrals. J Math Anal Appl 45:357–374
15. Davison ME (1981) A singular value decomposition for the Radon transform in n-dimensional Euclidean space. Numer Funct Anal Optim 3(3):321–340
16. Lisin FS (1976) Conditions for the completeness of the system of polynomials. Math Notes 18(4):891–894
17. Xu Y (2006) A new approach to the reconstruction of images from Radon projections. Adv Appl Math 36:388–420
18. Xu Y, Tischenko O, Hoeschen C. A new reconstruction algorithm for Radon data. Proceedings of SPIE, vol 6142. Medical imaging 2006: physics of medical imaging, pp 791–798
19. Hakopian H (1982) Multivariate divided differences and multivariate interpolation of Lagrange and Hermite type. J Approx Theory 34:286–305
20. Tischenko O, Xu Y, Hoeschen C (2010) Main features of the tomographic reconstruction algorithm OPED. Radiat Prot Dosimetry 139(1–3):204–207, Advance Access publication 12 Feb 2010
21. Noo F, Clackdoyle R, Pack JD (2004) A two-step Hilbert transform method for 2D image reconstruction. Phys Med Biol 49:3903–3023
22. Mikhlin SG (1957) Integral equations and their applications to certain problems in mechanics, mathematical physics and technology. Pergamon, New York
23. Tuy HK (1983) An inversion formula for cone-beam reconstruction. SIAM J Appl Math 43:546–552
24. Feldkamp LA, Davis LC, Kress JW (1984) Practical cone-beam algorithm. J Opt Soc Am A 1(6):612–619
25. Wang G, Lin T-H, Cheng PC (1993) A general cone-beam reconstruction algorithm. IEEE Trans Med Imaging 12:486–496
26. Turbell H (2001) Ph.D. Thesis

Chapter 7
Advances in SPECT Image Reconstruction

Gernot Ebel and Timo Aspelmeier

This chapter gives an overview of the recent developments in tomographic image reconstruction, focusing on those aspects which are particular to nuclear imaging. Every section of this chapter deals with a specific issue, which is known to be a limitation factor for the reconstruction, and describes the most recent strategies to overcome such problems.

7.1 Object Movement and Change

An essential requirement for any correct tomographic reconstruction is the stability of the measured object itself. This prerequisite has two aspects: the object should not move or cyclically morph and the object should not change its three-dimensional activity distribution over time.

The last aspect is rather specific for nuclear medicine. The spatial distribution of the incorporated activity must remain constant during the course of the measurement. All acquisition protocols therefore take care of the flow-in and the washout phase of the tracer. Sometimes however a change in tracer activity is unavoidable, for example, an accumulation in the bladder during a bone scan or a change of the activity pattern in the myocardium. As long as this change can be perceived as a locally defined monotonic increase or decrease of activity, it can be accounted for by so-called *dynamic* reconstruction [1]. More specifically, it is referred to as *5D* SPECT reconstruction for gated cardiac studies [2, 3]. Although this constraint makes the method very specific, the effect of artifact reduction is notable. In the currently available static cardiac tomographic SPECT cameras, the 5D approach finds its necessary hardware counterpart.

G. Ebel (✉) • T. Aspelmeier
Scivis wissenschaftliche Bildverarbeitung GmbH, 37085 Göttingen, Germany
e-mail: ebel@scivis.de

A. Giussani and C. Hoeschen (eds.), *Imaging in Nuclear Medicine*,
DOI 10.1007/978-3-642-31415-5_7, © Springer-Verlag Berlin Heidelberg 2013

Another aspect of the requirement of a stable object under measure is the absence of motion. When, however, the motion is an unavoidable aspect of the object itself, as it is true for the myocardium, this type of motion can quite successfully be accounted for by means of a gated acquisition, also referred to as 4D SPECT. Traditionally the reconstruction was performed separately for every gate, each of them corresponding to an ECG-sensed phase of the pulsating heart. As a matter of consequence, each reconstruction of a single gate suffered from noise as only a fraction of the whole myocardial cycle was acquired within a gate. Optical flow methods seem to be very promising attempts to increase the statistics of each gate, as they exploit the temporal continuity and periodicity of the motion field in the cyclic motion sequence. Although the work on this topic has been continuing for several years, proper reconstruction of discontinuities and object boundaries still represents a challenge for the researchers [4].

As the beating heart is embedded in a breathing thorax, two types of cyclic motion are effectively superimposed. The effect of taking into account the respiratory motion by respiratory gating acquisition and reconstruction is investigated and discussed in [5]. Both the independently acquired gated cardiac and respiratory motion can in principle be accounted for in the reconstruction. To make it complete, also the change of tracer distribution over the course of imaging can be included. However, the clinical feasibility of these efforts will be limited by the ability of the reconstruction scheme to deal with the tremendously increasing noise level of the double-gated projections.

7.2 Resolution: New Collimation Schemes

Traditionally, the well-established parallel hole collimators were accompanied by converging hole geometries such as the fan beam and the cone beam collimators. As the cone beam geometry introduces interdependence between neighboring transaxial planes, i.e., they cannot be reconstructed independently any more, this kind of collimation geometry did not find its way into clinical practice. With the raising computational power, however, the interest in the cone beam geometry was revived and is now used in a mixed setup, with parallel collimation at the edges of the field of view (FOV) and cone beam collimation at the center of the FOV, where the interesting clinical object is supposed to be. An example of this application is cardiac SPECT with proper patient positioning [6].

Preclinical imaging deals with small objects; therefore, good resolution images are required. The use of several pinholes as SPECT collimation (multi-pinhole collimation) has been well known for years and has its firm place in preclinical imaging as a means for high-resolution and high-sensitivity SPECT imaging. These multi-pinholes can be even so arranged that their exiting cones, facing towards the detector crystal, may overlap to a certain extent. As such, the tomographic FOV of the multi-pinhole collimation can be considered to be near the pinholes themselves. However, in clinical imaging, even if one focuses on brain and myocardial imaging, one cannot achieve a mean distance between pinhole and object of less than 10 cm. This is essential, as the detection sensitivity of a single pinhole drops off as the inverse cube of the distance between object and pinhole. The only way to circumvent the loss in sensitivity is to widen the pinhole diameter, which in consequence degrades the spatial resolution, in contrast to the reasoning that led to the idea of the multi-pinhole collimation in the first place (the gain in spatial resolution). What worsens the situation even more is the existence of regions with high tracer uptake near the FOV (such as the lobe of the liver in cardiac imaging). The emissions from these neighboring regions may enter into the tomographic field and lead to a degradation of the image. Nonetheless, there are some fruitful attempts employing multi-pinhole collimation for cardiac studies [7].

7.3 Quantification and Scatter

Compared with the impact caused by attenuation effects, the effects caused by scatter are far more subtle. Not taking into account the attenuation of the gamma quanta traveling from their origin through the patient towards the detector results in severely disproportioned reconstructed tracer distributions. This source of image degradation is to a large extent tamed by the advent of SPECT/CT systems and the corresponding use of their CT maps as attenuation prior for the SPECT reconstruction. Scatter correction, on the other hand, focuses on those detected events that originated from their source, also interacted with the patient's atoms, but were not absorbed as in the case of attenuation but were only deflected mainly by Compton interaction. If their loss of energy due to this interaction is small enough such that they still fall into the acquisition energy window, these events create a nonhomogeneous and object-dependent background that inhibits quantification.

In a thorough analysis of scatter corrections [8], those techniques trying to subtract the scatter events from the total recorded events are considered as rather crude, however, these are just the most prominently used ones at present. In addition, they are physically incorrect; they follow the dictum that *any* scatter correction is better than *no* scatter correction. The goal of quantitative SPECT however cannot be reached by these crude methods. For quantitative SPECT, one needs a physically correct consideration of the scatter, and the most straightforward method is to simulate the scattered events in the process of forward projection. This is done routinely by Monte Carlo (MC) methods and the prerequisite of a map of potential scatter centers can be transcribed from the CT map delivered by a SPECT/CT acquisition. The MC-based forward projection in its traditional way is however a problematic task for clinical settings, even with the computational power available today.

Some additional assumptions are needed to deliver the speedup in computation time that brings quantitative SPECT closer to clinical employment. First, the method of Forced Detection (FD) is a standard method for MC acceleration which considers only those scattered photons that travel towards the detector along the path that is dictated by ideal collimation. Second, it is possible to simulate the acceptance angle of the collimator bore holes by a convolution that is dependent on the distance between object and detector. The so-called Convolution FD (CFD) reduces the computational time by a factor of 50–100 as compared to FD and has been validated for ^{99m}Tc [9]. For other nuclides that are more prone to septal penetration, angular response functions are considered. The idea is the same as with the use of point spread functions (PSFs): to treat the object-dependent scatter separately from the apparatus-dependent influences which are constant.

7.4 Attenuation Correction by Simultaneous Emission and Transmission Map Reconstruction

Despite the fact that more and more newly sold SPECT camera systems are SPECT/CT systems, there is still ongoing research activity aiming to reconstruct both the SPECT emission map and the attenuation map solely on the basis of emission projections. As it is an obvious fact that the attenuation map leaves its mark on the measured emission projections, it is a kind of temptation to try to disentangle the different maps embedded in them.

Already around the year 2000, there were different ambitious approaches on this topic, some focusing on a postulated discrete consistency condition used to reduce the manifold of possible attenuation maps [10] and others employing nonlinear regularization approaches based on Tikhonov penalty terms used for parameter estimation problems [11]. Besides impressive results in well-chosen setups, none of them was finally employed routinely. New results step away from the attempt to calculate the attenuation map solely on the emission projections but take the instrumentally generated transmission map as a starting point for further refinements. Calling the measured attenuation map an attenuation prototype, this prototype is spatially transformed by nonlinear spline transformation models in order to take into account most prominently the breathing motion [12]. Of course, by properly registering any kind of patient motion, the attenuation map can be deformed accordingly, and the same treatment can be done on the emission map.

7.5 Correction for Partial Volume Effect

In ordinary situations, one cannot achieve a better spatial SPECT resolution than roughly 6 mm FWHM of a point source, even taking the distance-dependent collimator PSF into consideration. On the other hand, a voxel of a 128^3 pixel acquisition has a typical edge length of 3 mm. Therefore, the content of an ideal point source diffuses over many voxels. This underestimation of tracer activity, which becomes prominent as soon as the size of the true structure of the tracer distribution is roughly equal to or smaller than the reconstructed PSF, is called partial volume effect (PVE). By utilizing high-resolution structural images (CT or MRI), it is possible to compensate this effect. The major drawback is that one has to assume the morphologic region to be identical to the true tracer distribution. This causes partial volume correction (PVC) methods to not correct a single region only but also make possibly false assumptions about the tracer distributions.

Nevertheless, the field of PVC is still vivid, and the latest approaches tend to combine the PVC with the reconstruction process itself [13]. In order to assess different approaches and different reconstruction schemes that target the PVC, very recently a phantom was developed that employs continuously varying fillable cones. The cone diameter starts above the PVE region and decreases down into the PVE regime, and by coregistering the SPECT reconstructions with the CT (provided a SPECT/CT system is used), one can pinpoint the breakdown of the employed PVC [14]. More details on the phantom can be found in Chap. 10.

7.6 Theoretical Background: Compressive Sensing and Poisson Noise

It is of great interest to keep the SPECT measurement time as short as possible in order to reduce motion artifacts and to increase the patient's comfort. In addition, it is also desirable to reduce the administered activity to the lowest possible level. Both aims are hard to achieve because they reduce the number of photons counted, thus increasing the effect of quantum (i.e., Poisson) noise, or the number of angles viewed, or both. Traditional reconstruction methods have a hard time dealing with spatially incomplete and overly noisy data. Reconstruction problems are usually ill-posed, and this leads to an amplification of the noise in the measurements in the eventual reconstruction. A reduced number of angles make the problem even more ill-posed, and a reduced dose generates even higher Poisson noise; hence, the noise in the reconstruction becomes unacceptable. However, although every reconstruction method must employ some sort of prior information for regularization in order to tame the ill-posedness of the problem, a very powerful piece of information is usually ignored. This piece of information is that basically all real-world objects are sparse, or at least compressible, in a suitable representation. This means that the true information content of the object being reconstructed is not one number, the activity, for every voxel but in fact much less: in a suitable basis, only a few coefficients are actually needed in order to approximate the true object to a very good degree. In order to reconstruct these few coefficients, much less information must actually be measured, provided one knows how to extract that information from the measurements. This can be obtained employing the so-called compressive sensing (CS) techniques.

CS theory shows that a reconstruction which takes the sparsity information into account can be achieved by augmenting the usual maximum likelihood functional for Poisson statistics as it is used, for instance, in the MLEM algorithm and its many variants, by a sparsity-enforcing term which penalizes any non-null coefficients of the reconstruction in the sparse representation [15]. The activity distribution for which the minimum of this combined functional is attained is then the desired reconstruction.

As it turns out, there is a considerable problem with Poisson noise and compressive sensing. The nature of Poisson noise is such that the variance of the noise grows with the mean value. This is in contrast to the usual Gaussian noise which has a fixed variance, independent of the mean. The consequence is that the performance, i.e., the error of the eventual reconstruction, is not as good as for Gaussian noise. Nevertheless, it is still an improvement over conventional reconstruction.

A prerequisite for CS theory to work is that there must be a so-called incoherence between the sparsity basis and the measurement basis. Incoherence means that the basis vectors of the sparsity representation and the ones of the measurement representation must be as different from each other as possible. An example of maximal incoherence would be if the measurement basis were the identity basis, i.e., if a single measurement consisted of directly measuring the value of exactly one voxel, and the sparsity basis were the Fourier basis. In SPECT, this is obviously not the case: the measurement basis is simply the system matrix, i.e., one measurement consists of measuring the sum of photons which originate from any voxel on a line (or narrow cone) seen by a detector pixel. In addition, real-world objects are usually not sparse in Fourier space. However, good general sparsity bases exist in the form of wavelet bases.

In the CS literature, the measurement basis is often chosen to be a suitable random matrix which guarantees incoherence with very high probability. For SPECT, the system matrix is *fixed* and can thus not be chosen freely. The sparsity basis, for example, a suitable wavelet basis, can be chosen from a certain set of possible bases, but the choice is rather restricted: by far not every basis is a sparsity basis. Due to these restrictions, the incoherence is not optimal. Hence, the savings in dose and number of measurements cannot be as high as in other cases reported in the CS literature (e.g., transmission tomography techniques), but savings are nevertheless possible.

7.7 Acceleration: Implementation on GPUs

Nearly every topic mentioned in this section relies heavily on computational resources. The key technique today is the use of graphic cards for hardware acceleration. Only 5 years ago, one had to take cumbersome field programmable gate array (FPGA) implementations into consideration or the use of other, dedicated hardware. Nowadays, the multicore CPUs and the manifold of cores on a GPU are accessible by standard architectures such as Open CL and CUDA as extensions to C++ code. Still it is very problem-specific to find the fastest implementation; therefore, one cannot rely on former experiences, and it is as if every new implementation has to start from scratch. However, the tools to master the immensely grown need of computational resources are at our hands.

References

1. Wells RG, Farncombe T, Chang E, Nicholson RL (2004) Reducing bladder artifacts in clinical pelvic SPECT images. J Nucl Med 45:1309–1314
2. Feng B, Pretorius PH, Farncombe TH, Dahlberg ST, Narayanan MV, Wernick MN, Celler AM, Leppo JA, King MA (2006) Simultaneous assessment of cardiac perfusion and function using 5-dimensional imaging with Tc-99m teboroxime. J Nucl Cardiol 13:354–361
3. Xiaofeng N, Yongyi Y, Mingwu J, Wernick MN, King MA (2010) Regularized fully 5D reconstruction of cardiac gated dynamic SPECT images. IEEE Trans Nucl Sci 57:1085–1095
4. Wenyuan Q, Xiaofeng N, Yongyi Y (2011) An improved periodic optical flow model for cardiac gated image reconstruction. In: Biomedical imaging: from nano to macro. IEEE international symposium, pp 1276–1279
5. Segars WP, Mok SP, Tsui B (2009) Investigation of respiratory gating in quantitative myocardial SPECT. IEEE Trans Nucl Sci 56:91–96
6. Newsfeed of medicalphysicsweb: Siemens demonstrates IQ.SPECT for ultrafast nuclear cardiology. http://www.medicalphysicsweb.org/cws/article/newsfeed/45619
7. Garcia EV, Faber TL, Esteves FP (2011) Cardiac dedicated ultrafast SPECT cameras: new designs and clinical implications. J Null Med 52:210–217
8. Buvat I, Benali H, Todd-Pokropek A, Di Paola R (1994) Scatter correction in scintigraphy: the state of the art. Eur J Nucl Med 21:675–694
9. De Beenhouwer J, Staelens S, Vandeberghe S, Lemahieu I (2008) Acceleration of GATE SPECT simulations. Med Phys 35:1476–1485
10. Bronnikov AV (2000) Reconstruction of attenuation map using discrete consistency conditions. IEEE Trans Med Imaging 19:451–462
11. Dicken V (1999) A new approach towards simultaneous activity and attenuation reconstruction in emission tomography. Inverse Probl 15:931–960
12. Barendt S, Modersitzki J (2011) SPECT reconstruction with a non-linear transformed attenuation prototype. In: Handels H et al (eds) Bildverarbeitung für die Medizin 2011, Informatik aktuell. Springer, Berlin, pp 414–418
13. Erlandsson K, Thomas B, Dickson J, Hutton BF (2011) Partial volume correction in SPECT reconstruction with OSEM. Nucl Instrum Method Phys Res A 648(Suppl 1):S85–S898
14. Engeland U (2010) Phantom for a tomographic medical imaging apparatus. European Patent application publication number EP2 267 484
15. Aspelmeier T, Ebel G, Hoeschen Ch (2011) Tomographic imaging using Poissonian detector data. International patent application publication number WO 2011/091815

Chapter 8
Noise Reduction

Oleg Tischenko and Christoph Hoeschen

In this chapter, we will describe the theoretical background of noise reduction in medical imaging as well as give some examples of noise reduction methods. To do so, we start with a fundamental description of digital image generation in medical imaging, since we will only focus on digital images and noise reduction by means of digital image processing. In the next part, we will discuss the corresponding processing in general before we describe the approaches typically used mainly based on linear filtering and some new approaches based on nonlinear approaches.

8.1 Idealized Model of Digital Image Generation

The simplified model of an imaging system can be characterized by the image space $\mathbf{D}$, which can be assumed to be two dimensional without a loss of generality, and impulse response φ. Let $\mathbf{D} \subset \mathbb{R}^2$ be squared area of the size $D \times D$, and let φ be a continuous function of fast decay whose Fourier transform $\hat{\varphi}(\omega_x, \omega_y)$ is supported inside the circle $\omega_x^2 + \omega_y^2 = \Omega_0^2$, i.e., $\hat{\varphi}(\omega_1, \omega_2) = 0$ if $\omega_1^2 + \omega_2^2 > \Omega_0^2$. The output $F : \mathbf{D} \to \mathbb{R}$ of the imaging system is modeled as a convolution

$$F = f * \varphi, \tag{8.1}$$

where $f : \mathbf{D} \to \mathbb{R}$ is the input signal. It is supposed that f is integrable on $\mathbf{D}$. Then F can be represented by the Fourier series

O. Tischenko (✉) • C. Hoeschen
Research Unit Medical Radiation Physics and Diagnostics, Helmholtz Zentrum München - German Research Center for Environmental Health, Neuherberg, Germany
e-mail: oleg.tischenko@helmholtz-muenchen.de

A. Giussani and C. Hoeschen (eds.), *Imaging in Nuclear Medicine*,
DOI 10.1007/978-3-642-31415-5_8, © Springer-Verlag Berlin Heidelberg 2013

$$F(x, y) = \sum_{m, k \in \mathbb{Z}} \hat{F}_{m, k} e^{i\delta(xm+yk)}, \tag{8.2}$$

where $\hat{F}_{m,k} = D^{-2}\hat{f}(\delta m, \delta k)\hat{\varphi}(\delta m, \delta k)$ and

$$\delta = 2\pi/D. \tag{8.3}$$

Since the impulse response φ is band limited with the bandwidth Ω_0, the series (8.2) can be truncated:

$$F(x,y) = \sum_{-N \leq m,k \leq N} \hat{F}_{m,k} e^{i(xm+yk)\delta} \tag{8.4}$$

with

$$N = \left\lceil \frac{D}{\Delta} \right\rceil \tag{8.5}$$

and

$$\Delta = \frac{2\pi}{\Omega_0} \tag{8.6}$$

That is, the output F is a trigonometric polynomial and, consequently, can be uniquely recovered from its samples (see, e.g., [1]) measured over the rectangular grid $\{(x_m, y_k)|x_m = m\Delta, y_k = k\Delta\}_{-N \leq m,k \leq N}$. Hence, under certain idealizing assumptions, the continuous image can be identified with its samples.

We refer to the matrix $\{F_{m,k}\}_{-N \leq m,k \leq N}$ as a digital image of the input signal f. In the following, we call the number N the *resolution limit* of the imaging system, meaning that $N\delta$ is the maximal resolution that can be achieved with this system. Clearly, for the case considered here, the value $N\delta$ coincides with the Nyquist frequency of the imaging system.[1]

Finally, we point out that extending the matrix $\{F_{j,n}\}$ periodically with the period $M \times M$, $M = 2N + 1$, we can write

$$\hat{F}_{l,m} = \frac{1}{M^2} \sum_{j=0}^{M-1} \sum_{k=0}^{M-1} F_{j,k} e^{-2\pi i(jl+km)/M}, \tag{8.7}$$

which is the standard notation for the discrete 2D Fourier transform.

[1] It is not always the case in real systems.

8.2 Image Processing

In practice, we do not have to deal with the image F but with its corrupted version I

$$I = F + \eta, \tag{8.8}$$

where η is the random value referred to as system noise. For an X-ray imaging system, another substantial noise component, which is due to the Poisson statistics and scattering of the X-ray quanta, is referred to as quantum noise (see, e.g., Chap. 9 of [2]). In view of the fact that the quantum noise is an inherit feature of the input signal f, the representation (8.8) can be replaced by

$$I = (g + q) * \varphi + \eta, \tag{8.9}$$

where q is the input quantum noise and g is the *desired* signal. One of the most important problems of the image processing is to recover g from the measured image I. It is clear, that this can be done to some degree of accuracy only. The best possible approximation of g in terms of the mean squared error is given by

$$Ag(x,y) = \sum_{-N \le k,l \le N} \hat{g}_{k,l} \mathrm{e}^{\mathrm{i}(xk+yl)\delta}, \tag{8.10}$$

where, at the right, there is the Fourier series of g truncated up to the resolution limit of the given imaging system. Leaving aside details, we mention that the linear problem of recovering unknowns $\hat{g}_{k,l}$ from I is sometimes referred to as the Wiener filtering.[2]

As it follows from (8.9), the problem of recovering Ag from I can be looked at as the one consisting of two distinct tasks: the suppression of the noisy component $q * \varphi + \eta$ (*noise reduction*) and the deconvolution of blurred image $g * \varphi$ (*de-blurring*). In the following, we use this discrimination in two tasks and describe some approaches used for the reduction of noise in the image.

8.3 Linear Filtering

The general strategy of this approach is to separate in the spectral description of the medical image the frequencies corresponding mainly to signals and those mainly corresponding to noise. Afterwards it is tried to suppress the power of those spectral components which are related to noise. This is done via appropriate weighting of

[2] Norbert Wiener had stated and solved this problem under special conditions for stationary stochastic time series (see [3]).

the frequency components of image I. The result of the weighting is the image wI, given by

$$wI(x,y) = D^{-2} \sum_{-N\leq l,m\leq N} \hat{w}_{l,m} \hat{I}_{l,m}\, \mathrm{e}^{\mathrm{i}(xl+ym)\delta}. \tag{8.11}$$

The set $\{\hat{w}_{l,m}\}_{-N\leq l,m\leq N}$ is called transfer matrix. The elements of transfer matrix are samples of the Fourier transform of the filter w, in terms of which the right-hand side of (8.11) can be written as the 2D convolution

$$wI = I * w. \tag{8.12}$$

In the following, we call function $w \in L_2(\mathbb{R}^2)$ *smoothing function* if it is continuous and

$$\hat{w}(0) = \int_{\mathbb{R}^2} w(x,y)\mathrm{d}x\mathrm{d}y \neq 0. \tag{8.13}$$

Due to (8.12) the kernel w acts upon the image I as a local weighted averaging, which is uniform for the whole image and is independent from its current space context. Linear filtering is perfect if the power spectrum of the image I can be separated in two disjoint regions, the one of which contains the spectral component of the useful signal and, the other, of the noise. We illustrate this apparent property in Fig. 8.1, where in the middle, there is an image corrupted with additive noise, the spectral power of which is concentrated outside the circle $\omega_x^2 + \omega_y^2 = \Omega^2$. On the contrary, the spectral power of the useful signal (left) is concentrated inside this circle. The right image is the convolution of the corrupted image with the ideal filter, i.e., the filter, the transfer function of which is

$$\hat{w}(\omega) = \begin{cases} 1, & |\omega| \leq \Omega, \\ 0, & \text{otherwise}. \end{cases} \tag{8.14}$$

Clearly, in this case, the original image and the denoised one are identical. In fact, if the power spectrum of the useful signal is concentrated inside a bounded region of the frequency domain and the signal-to-noise ratio is big enough in this region, then the linear filtering is still excellent independently from the spectral distribution of the noise. The corresponding example is illustrated in Fig. 8.2. On the left of this figure, there is the same image as in Fig. 8.1 with added white noise and, on the right, the result of the filtering with Gaussian filter g_Ω, the transfer function of which is

$$\hat{g}_\Omega(\omega) = \mathrm{e}^{-\frac{1}{2}\left(\frac{|\omega|}{\Omega}\right)^2}. \tag{8.15}$$

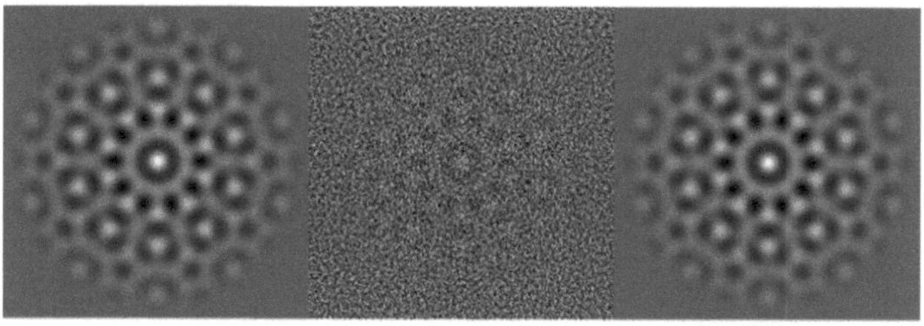

Fig. 8.1 Linear filtering of the band-limited image. *Left*: the image the power spectrum of which is inside the circle B. *Middle*: the same image with added noise, the power spectrum of which is outside the circle B. *Right*: convolution of the middle image with the ideal filter given by (8.14)

Fig. 8.2 Linear filtering of the band-limited image. *Left*: image from Fig. 8.1 with added white noise. *Right*: the convolution with Gaussian kernel

Thus, the following short conclusion can characterize the linear filtering: it works well if applied to regular signals, i.e., such signals the power spectrum of which is concentrated within a bounded region of low frequencies and it vanishes fast outside this region. The filtering operation in this case is reduced to the suppression of high-frequency components of the image. If the Fourier transform of an image decays slowly as $|\omega| \to \infty$, the linear filtering is not efficient anymore. Since the spectral power of such images is high also in the high-frequency sub-band, suppressing within this sub-band affects the sharpness of edges and destroys fine details of the image. This results in overall blurring of the image (see Fig. 8.3).

There is a variety of filters and corresponding smoothing kernels which are available. However, all rely on the same principle and show similar behavior.

8.4 Adaptive Nonlinear Filtering

In contrast to methods based on linear filtering, many advanced noise reduction techniques assume a control over the local amount of smoothing. Appropriately choosing a smoothing kernel at the current positions, it is possible to avoid unnecessary blurring in regions of high variation of the useful signal. Usually in practice the desired kernel is chosen from a predefined family of kernels. As an example we consider the family $\{g_\sigma, \sigma \geq 0\}$, where

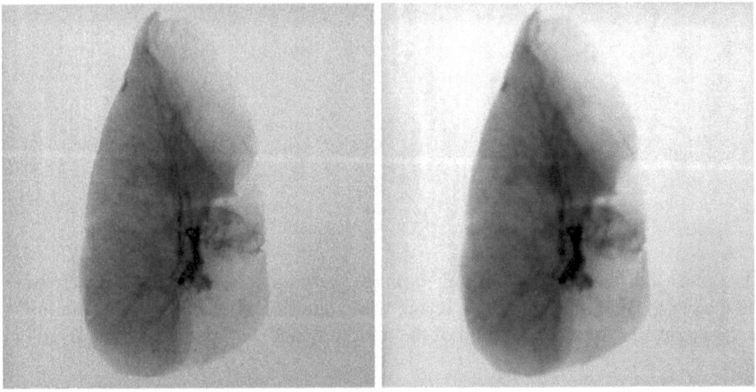

Fig. 8.3 Blurring effect of the linear filtering. *Left*: X-ray image of a human lung specimen. *Right*: convolution with a Gaussian kernel

$$g_\sigma(x,y) = \sigma^{-2} g\left\{\frac{x}{\sigma}, \frac{y}{\sigma}\right\} \tag{8.16}$$

and the *generating kernel* g is a smoothing kernel [see definition above, (8.13)]. It can easily be seen that for $\sigma > 0$, $\hat{g}_\sigma(0) = 1$, that is, g_σ is also a smoothing kernel. Besides, $\lim\limits_{\sigma\to 0} F * g_\sigma(x,y) = F(x,y)$, which implies that $\lim\limits_{\sigma\to 0} g_\sigma = \delta$, where δ is a Dirac delta function. These properties allow to construct the image gF defined by

$$gF(x,y) = \int F(t,s) g_{\sigma(x,y)}(x-t, y-s)\mathrm{d}t\mathrm{d}s, \tag{8.17}$$

where we determine σ depending on current position (x, y). The challenging task is to establish a rule according to which the value σ can be adapted to the current context of the image. Usually, σ is required to change as a function of some *decisive characteristic* such as the gradient, the Laplacian, the curvature, or some other local characteristic of the image. An example of such filtering with the Gaussian function

$$g(x,y) = \frac{1}{2\pi} \mathrm{e}^{-\frac{x^2+y^2}{2}} \tag{8.18}$$

as a generating kernel, and the squared deviation $\xi(x,y)$ as a local decisive characteristic, is given in Fig. 8.4. Particularly for this example, the current value of σ was set to

$$\sigma(x,y) = h\left(C; \frac{\xi_r(x,y)}{\xi_{\max}}\right), \tag{8.19}$$

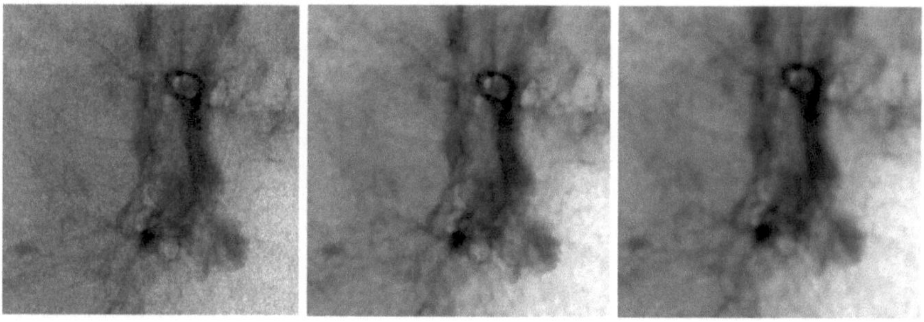

Fig. 8.4 Filtering with adaptive bandwidth Gaussian kernel. *Left*: a patch of the X-ray image shown in Fig. 8.3; *middle*: the result of the filtering by given tuning parameters r and $C = 5r$; *right*: linear filtering with Gaussian kernel by $\sigma = r$

where

$$h(C;t) = \frac{r}{1 + Ch_1(t)} \tag{8.20}$$

and the function $h_1 : [0, 1] \to [0, 1]$ is monotonically increasing from zero to one. For the local squared deviation, we have

$$\xi_r^2(x,y) = \frac{1}{2\pi r^2} \int_{B_r(x,y)} |F(t,s) - \bar{F}(x,y)|^2 \mathrm{d}t\mathrm{d}s, \tag{8.21}$$

$$\xi_{\max} = \max_{(x,y)\in\mathbf{D}} \xi_r(x,y), \tag{8.22}$$

where the ball $B_r(x,y)$ is a ball of radius r located in (x,y); $\bar{F}(x,y)$ is the local mean value of F inside the ball.

The smoothing considered in this example incorporates two tuning parameters, which are the constants C and r. From (8.19) it follows that σ varies between r and $r/(C+1)$, and the type of this variation depends on the type of the growth of the function h_1 defined in (8.20).

Another class of smoothing techniques is the so-called sigma filtering. Here we use smoothing kernels of the following general form:

$$\varphi_{I;x,y}(t,s) = C(x,y)w_\tau(I(t,s) - I(x,y))g_\sigma(x-t, y-s), \tag{8.23}$$

where (x,y) is the current *update position*; g_σ and w_τ are the smoothing kernels, the effective width of which are σ and τ, respectively; and the local constant $C(x,y)$ is determined from the condition

$$\int_{\mathbb{R}^2} \varphi_{I;x,y}(t,\, s)\mathrm{d}t\mathrm{d}s = 1. \tag{8.24}$$

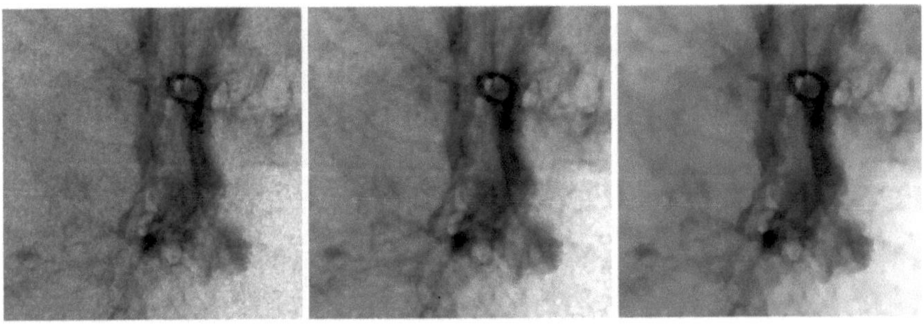

Fig. 8.5 Sigma filtering. Three iterations of applying sigma filter to the patch of the lung phantom

Because of two kernels acting both in the space and the *intensity* domains, sigma filtering is sometimes referred to as *bilateral* filtering (see [4]). As it follows from (8.23), the sigma filter is represented by a family of kernels which depends on two scale parameters σ and τ. The parameter σ determines the radius of the ball $B(x, y)$ centered in (x, y). Using the parameter τ, it has to be decided which values $I(x - t, y - s)$ inside the ball $B(x, y)$ have to be weighted down. Normally, these are values which deviatc from the value $I(x, y)$. In other words, the kernel w generates the local *averaging mask* that allocates those points within the ball $B(x, y)$ where image values are close to $I(x, y)$.

Bilateral filters are especially efficient if applied iteratively. Denoting the filtering operation with F, the iterative chain can formally be represented as $\mathrm{F}_{\sigma_n \tau_n} \mathrm{F}_{\sigma_{n-1} \tau_{n-1}} \ldots \mathrm{F}_{\sigma_1 \tau_1}$.

In [5], it has been shown that the best result by an iterative sigma filter can be achieved if $\sigma_1 > \ldots > \sigma_{n-1} > \sigma_n$ and $\tau_1 < \ldots < \tau_{n-1} < \tau_n$. It can be shown that such a sequence of filtered images converges to the image which is piecewise constant. As an example, in Fig. 8.5, there are three iterations of bilateral filtering where w and g are both Gaussian.

8.5 Wavelet-Based Nonlinear Filtering

A quite different principle for spatially adaptive noise reduction is based on the reconstruction from selected wavelet coefficients. For the sake of completeness, before describing the related techniques, we will review main points of the theory of the wavelet transform referring mainly to [6] and [7]. In doing so, we will use the terminology of the theory of functions, and therefore, we start by stating formal notations accepted in this theory and used in the text.

C_m, $m \geq 1$, is a space whose elements are m times continuously differentiable functions; the space C_0 is the space of continuous functions, and C_{-1}, the space of piecewise continuous functions.

Let us have $j \in \mathbb{Z}$ and $m \in \mathbb{N}$. As *spline space* S_m^j, we will call the set of all such functions $f \in C_{m-2}$, the restriction of which to the interval $[2^j k, 2^j(k+1)]$ is the algebraic polynomial of order $m-1$; the points $2^j k$ are called knots of the spline space S_m^j.

A space with an inner product $\langle \cdot, \cdot \rangle$ is called Hilbert space. In the following, the norm of element $a \in H$ is defined by

$$\|a\|_H = \sqrt{\langle a, a \rangle}. \tag{8.25}$$

$L_p(T)$ is a space of functions satisfying $\int_T |f(x)|^p < \infty$. The space $L_2(T)$ is a Hilbert space with the inner product

$$\langle a, b \rangle = \int_T a(x) b(x)^* \mathrm{d}x, \tag{8.26}$$

where b^* is a complex conjugate of b.

ℓ_p is a set of all infinite sequences $\{a_k\}_{k \in \mathbb{Z}}$ such that $\sum_k |a_k|^p < \infty$. The space ℓ_2 is a Hilbert space with the inner product

$$\langle a, b \rangle = \sum_k a_k b_k^*. \tag{8.27}$$

Span $\{g_n\}_n$ is the minimal space *spanned* on the family $\{g_n\}_n$, i.e., for any $a \in \ell_2$,

$$\sum\nolimits_n a_n g_n = f \in \operatorname{span}\{g_n\}_n. \tag{8.28}$$

Operators that map a Hilbert space E into Hilbert space G will be denoted as $\mathbf{H} : E \to G$. With $\mathbf{H}^*$, we denote the adjoint of $\mathbf{H}$, i.e., such that $\langle \mathbf{H}/f, g \rangle_G = \langle f, \mathbf{H}^* g \rangle_E$ for any $f \in E$ and any $g \in G$. The operator is self-adjoint, if $\mathbf{H} = \mathbf{H}^*$.

A wavelet is a function $\psi \in L_2(\mathbb{R})$ that necessarily satisfies the condition

$$\int_{\mathbb{R}} \psi(x) \mathrm{d}x = \hat{\psi}(0) = 0. \tag{8.29}$$

Everywhere in the text, it will be supposed that ψ is real and normalized, i.e., $\|\psi\|_{L_2} = 1$. Wavelets are "famous" for being well localized both in the space domain and in the frequency domain. The localization properties of wavelets are usually expressed in terms of a so-called *space-frequency window*. This is a rectangle $\sigma \times \hat{\sigma}$ in the space-frequency plane with σ and $\hat{\sigma}$ defined by

$$\sigma^2 = \int_{\mathbb{R}} (x - \bar{x}) |\psi(x)|^2 \mathrm{d}x, \quad \bar{x} = \int_{\mathbb{R}} x |\psi(x)|^2 \mathrm{d}x, \tag{8.30}$$

$$\hat{\sigma}^2 = \pi^{-1} \int_0^\infty (\omega - \bar{\omega})^2 |\hat{\psi}(\omega)|^2 \mathrm{d}x, \quad \bar{\omega} = \pi^{-1} \int_0^\infty \omega |\hat{\psi}(\omega)|^2 \mathrm{d}x. \tag{8.31}$$

The point $(\bar{x}, \bar{\omega})$ is the center of the space-frequency window of the wavelet. Since ψ is real, the modulus of its Fourier transform is even, and this is why the integration in (8.31) is made over the positive half axis only.

Wavelet ψ is called the *mother wavelet* for the family $\{\psi_{s,u}\}_{s>0, u\in\mathbb{R}}$, where the function

$$\psi_{s,u}(x) = s^{-1/2} \psi\left(\frac{x-u}{s}\right) \tag{8.32}$$

is the dilated and translated version of the mother wavelet ψ. As an example, let us consider the family generated by so-called Mexican hat wavelet

$$\lambda(x) = \frac{2}{\sqrt{3}} \pi^{-1/4} (1 - x^2) \mathrm{e}^{-\frac{x^2}{2}}. \tag{8.33}$$

In Fig. 8.6 wavelets $\lambda_{1,0}$ and $\lambda_{5,0}$ of this family (left) as well as their Fourier transforms (right) are shown.

Figure 8.7 shows space-frequency windows of these wavelets. One observes that the bigger is s, the wider is the effective width of the wavelet in the space domain and the better is its spectral resolution. In contrary, we obtain better space resolution by smaller s. Since $\hat{\sigma}_s = \hat{\sigma}/s$ and $\sigma_s = s\sigma$, the area of the space-frequency window does not change.

The bivariate function

$$wf(s, u) = \langle f, \psi_{s,u} \rangle, \ s, \ u \in \mathbb{R} \tag{8.34}$$

is called a continuous wavelet transform of the function f. Since

$$\langle \hat{f}, \hat{\psi}_{s,u} \rangle = 2\pi \langle f, \psi_{s,u} \rangle, \tag{8.35}$$

the wavelet coefficient $\langle f, \psi_{s,u} \rangle$ by fixed s, u relates to the local contribution made by f within the space-frequency window of the wavelet $\psi_{s,u}$.

Besides the space-frequency window, another important characteristic of a wavelet is the number of vanishing moments. A wavelet ψ is said to have m vanishing moments if $\int x^k \psi(x) \mathrm{d}x = 0$ for $0 \leq k < m$. Since

$$\frac{\mathrm{d}^n}{\mathrm{d}\omega^n} \hat{\psi}(\omega) = \int (-ix)^n \psi(x) \mathrm{e}^{-i\omega x} \mathrm{d}x, \tag{8.36}$$

the number of vanishing moments of ψ is equal to the number of zeros of its Fourier transform at $\omega = 0$. Using this property, it can be shown that a compactly supported

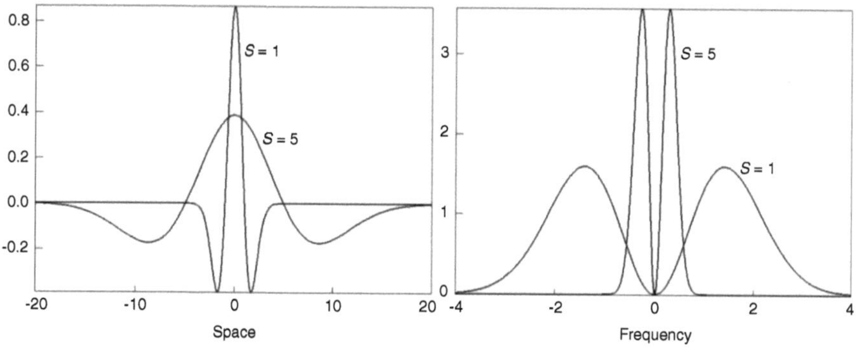

Fig. 8.6 Mexican hat wavelet. Wavelets $\lambda_{s,0}$ for $s = 1, 5$ (*left*) and their Fourier transforms (*right*)

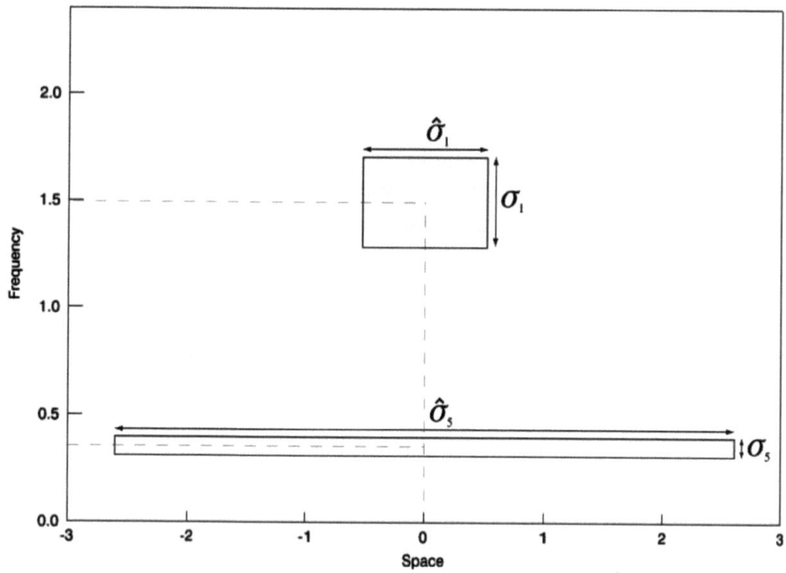

Fig. 8.7 Space-frequency windows of $\lambda_{s,0}$ for $s = 1, 5$

wavelet ψ has m vanishing moments if and only if there exists a compactly supported smoothing function θ such that

$$\psi(t) = (-1)^n \frac{\mathrm{d}^n \theta(t)}{\mathrm{d}t^n} \tag{8.37}$$

(see Chap. 6 of [7] for details); the corresponding wavelet transform of f is consequently

$$Wf(s,t) = s^{n+1/2} \frac{\mathrm{d}^n}{\mathrm{d}t^n} (f * \bar{\theta}_s), \tag{8.38}$$

where $\bar{\theta}_s(t) = \theta_s(-t)$ and $\theta_s(t) = s^{-1}\theta(t/s)$. Hence, the number of vanishing moments is crucial in those applications where the local regularity of the signal has to be measured.

The identity

$$f(x) = C_\psi^{-1} \int_0^\infty \int_{-\infty}^\infty \langle f, \psi_{s,u} \rangle \psi_{s,u}(x) \mathrm{d}u \frac{\mathrm{d}s}{s^2} \tag{8.39}$$

with

$$C_\psi = \int_0^\infty \frac{|\hat{\psi}(\omega)|^2}{\omega} \mathrm{d}\omega \tag{8.40}$$

is valid and makes sense for any $f \in L_2\ \mathbb{R}$ if $C_\psi < \infty$. Therefore, the wavelet ψ is called admissible if $C_\psi < \infty$. Condition (8.29) formulated above is necessary for ψ to be admissible. However, assuming that ψ is "reasonable," that is, ψ is such that it makes sense in the practice, this condition is also sufficient.

There are different possibilities to choose wavelets in 2D. For example, this can be a wavelet $\psi \in L_2(\mathbb{R}^2)$ that is radially symmetric. Since the wavelet in this case depends effectively on one variable only, the situation is essentially the same as in the 1D case: the 2D wavelet transform is the function that depends on two parameters, the shift (which is now 2D vector) and the dilation. The reconstruction formula is in this case

$$f(\mathbf{r}) \sim \int_0^\infty \int_{\mathbb{R}^2} \langle f, \psi_{s,\mathbf{u}} \rangle \psi_{s,\mathbf{u}}(\mathbf{r}) \mathrm{d}\mathbf{u} \frac{\mathrm{d}\mathbf{s}}{\mathbf{s}^3}. \tag{8.41}$$

It is also possible to choose a 2D wavelet that is not radially symmetric. Such wavelets are often called oriented wavelets. The wavelet transform in this case is a function that depends on three parameters, dilation, translation, and rotation, and the corresponding reconstruction formula is

$$f(\mathbf{r}) \sim \int_0^{2\pi} \int_{\mathbb{R}^2} \int_0^\infty \langle f, \psi_{s,\mathbf{u},\theta}, \rangle \psi_{s,\mathbf{u},\theta}(\mathbf{r}) \mathrm{d}\mathbf{u} \frac{\mathrm{d}\mathbf{s}}{\mathbf{s}^3} \mathrm{d}\theta. \tag{8.42}$$

In practice, one often applies separable 2D wavelets, i.e., a wavelet of the form $\varphi(t)\psi(s)$, where both φ and ψ are functions from $L_2(\mathbb{R})$, and at least the one of which is a wavelet.

The continuous wavelet transform provides an extremely redundant description of the function f. In fact, even a discrete subset of $\{Wf(s,u)\}_{s,u}$ can be used to recover functions exactly. A related discrete wavelet transform is $Wf_{m,n} = \langle f, \psi_{m,n} \rangle$ where

$$\psi_{m,n}(x) = a^{-m/2}\psi(a^{-m}x - nb) \tag{8.43}$$

for some fixed positive a, b.

For a wide range of choices for ψ, a and b discrete families of wavelets constitute *frames* of the functional space $L_2(\mathbb{R})$. The family $\{g_n\}_n$ is called a frame of the Hilbert space H if there exist constants $0 < A \le B < \infty$ such that

$$A\|f\|_H^2 \le \sum_n |\langle f, g_n\rangle|^2 \le B\|f\|_H^2 \tag{8.44}$$

for any $f \in H$. The equivalency $\langle f, g_n\rangle \equiv 0 \Leftrightarrow f = 0$ that follows from this definition implies that frames are complete in H, which means that any function in H can be approximated by linear combination of frame elements with any degree of accuracy. The frame is called redundant if its elements are linearly dependent.

In general, recovering a function from the frame coefficients is equivalent to the inversion of a self-adjoint operator $\mathbf{H} : H \to H$ that relates to the frame $\{g_n\}_n$ of H via

$$\mathbf{H}f = \sum_n \langle f, g_n\rangle g_n. \tag{8.45}$$

It is easy to see that the frame condition (8.44) is equivalent to the bounding condition

$$A\|f\|^2 \le |\langle \mathbf{H}f, f\rangle| \le B\|f\|^2 \tag{8.46}$$

that holds for any $f \in H$. This means that the inverse $\mathbf{H}^{-1}$ exists and is stable. Applying $\mathbf{H}^{-1}$ to both sides of (8.45) yields

$$f = \sum_n \langle f, g_n\rangle \tilde{g}_n \tag{8.47}$$

with

$$\tilde{g}_n = \mathbf{H}^{-1} g_n. \tag{8.48}$$

The family $\{\tilde{g}_n\}_n$ is called dual frame. Indeed, let us introduce the operator $\tilde{\mathbf{H}}$ that relates to $\{\tilde{g}_n\}_n$ via

$$\tilde{\mathbf{H}}f = \sum_n \langle f, \tilde{g}_n\rangle \tilde{g}_n. \tag{8.49}$$

One directly verifies that $\tilde{\mathbf{H}} = \mathbf{H}^{-1}$, and since $\mathbf{H}^{-1}$ is bounded by B^{-1} and A^{-1}, $\{\tilde{g}_n\}$ satisfies the frame bounding conditions with the bounds B^{-1} and A^{-1}. Moreover, the dual of $\{\tilde{g}_n\}_n$ is $\{g_n\}_n$, and therefore,

$$f = \sum_n \langle f, \tilde{g}_n \rangle g_n. \tag{8.50}$$

The inverse of $\mathbf{H}$ can be determined by means of the auxiliary operator $\mathbf{R} = \mathbf{I} - 2(A+B)^{-1}\mathbf{H}$ where $\mathbf{I}$ is the identity operator. The operator $\mathbf{R}$ is self-adjoint and satisfies the bounding condition

$$|\langle \mathbf{R}f, f \rangle| \leq \frac{B-A}{A+B} \|f\|_{L^2}^2 \leq \|f\|_{L^2}^2 \tag{8.51}$$

which means that $\|\mathbf{R}\| < 1$. Then,

$$\mathbf{H}^{-1} = 2(A+B)^{-1}(I - \mathbf{R})^{-1} = 2(A+B)^{-1} \sum_{k=0}^{\infty} \mathbf{R}^k. \tag{8.52}$$

In practice, one truncates the series (8.52) up to some N and uses the approximation

$$\tilde{g}_n \approx \tilde{g}_n^N = 2(A+B)^{-1} \sum_{k=0}^{N} \mathbf{R}^k g_n. \tag{8.53}$$

It is known (see Chap. 3 of [6]) that

$$\left\| f - \sum_n \langle f, g_n \rangle \tilde{g}_n^N \right\|_{L^2} \leq \left(\frac{B-A}{A+B} \right)^{N+1} \|f\|_{L^2}. \tag{8.54}$$

That is, the norm in (8.54) converges exponentially to zero at a rate depending on the value $(B-A)/(A+B)$. If $A \approx B$, one can truncate the series (8.52) up to zeroth term avoiding the computation of the dual frame and still have a high quality of reconstruction of arbitrary function.

If $A = B$, the frame is called *tight*. In this case, we can rearrange the identity

$$\sum_n |\langle f + g, g_n \rangle|^2 = A\|f + g\|_H^2 \tag{8.55}$$

to the identity

$$\left\langle f - A^{-1} \sum_n \langle f, g_n \rangle g_n, h \right\rangle = 0 \tag{8.56}$$

that holds for any $h \in H$ and any $f \in H$, and as a result, obtain the inversion formula

$$f = A^{-1} \sum_n \langle f, g_n \rangle g_n \tag{8.57}$$

that holds for any $f \in H$ at least in the weak sense.[3] The simplicity of (8.57) makes tight frames especially interesting in practice. Note that in spite of the similarity between (8.57) and a decomposition in an orthonormal basis, tight frames can be redundant. In the following, we show that a tight frame is either redundant or orthonormal. Indeed, setting $f = g_k$ in (8.56), we obtain that for any k, $\sum_n \alpha_n g_n = 0$ where

$$\alpha_n = \begin{cases} A^{-1}\langle g_k, g_k \rangle - 1, & n = k, \\ A^{-1}\langle g_n, g_k \rangle, & n \neq k.. \end{cases} \tag{8.58}$$

This means that if $\{g_n\}$ are linearly independent, then $\langle g_n, g_k \rangle = A\delta_{n,k}$, i.e., $\{g_n\}_n$ is orthonormal. On the other side, if $\|g\|_H^2 = A$ for all n, then

$$\sum_n |\langle g_k, g_n \rangle|^2 = A^2 + \sum_{n \neq k} |\langle g_k, g_n \rangle|^2 = A^2, \tag{8.59}$$

which is possible only if $\langle g_k, g_n \rangle = A\delta_{k,n}$. It follows that if for some n, $\|g_n\|_H^2 = C \neq A$, then the tight frame is redundant and $C \leq A$. For a tight frame, elements of which are of the same norm, the value A/C, where $\|g_n\|_H^2 = C$, can be interpreted as a measure of the redundancy of the frame.

Redundant tight frames can be used for noise reduction. Suppose that the signal f is measured over a redundant frame $\{g_n\}_n$, and let the measured coefficients f_n be contaminated with noise, i.e., $f_n = \langle f, g_n \rangle + q_n$, where q is a random variable. Let $\mathbf{U} : H \to \ell_2$ be the *frame operator*

$$(\mathbf{U}f)_n = \langle f, g_n \rangle. \tag{8.60}$$

Since the frame is redundant, there exists nontrivial $c \in \ell_2$, so that $\sum_n c_n g_n = 0$. In other words, the orthogonal complement of $\operatorname{Im} \mathbf{U}$ in ℓ_2, where we denote the image of the operator $\mathbf{U}$ as $\operatorname{Im} \mathbf{U}$, is not empty. Let $\mathbf{P} : \ell_2 \to \ell_2$ be the orthoprojector onto $\operatorname{Im} \mathbf{U}$. Then, $\mathbf{P}(\mathbf{U}f + q) = \mathbf{U}f + \mathbf{P}q$. Decomposing $q = q_1 + q_2$, where $q_1 \in \operatorname{Im}$, and $q_2 \in (\operatorname{Im} \mathbf{U})^\perp$, where $(\operatorname{Im} \mathbf{U})^\perp$ is the orthogonal complement of $\operatorname{Im} \mathbf{U}$ in ℓ_2, we obtain that $\mathbf{P}q = q_1$, which in turn implies that $\|\mathbf{P}q\|_{\ell_2} \leq \|q\|_{\ell_2}$. The following result is due to [7, Chap. 5]: let $\|g_n\|_{L_2}^2 = C$ for all n, and let q be a zero-mean white noise of variance σ_q^2. Then,

[3] The equality $f = g$ holds in the weak sense if $\langle f, h \rangle = \langle g, h \rangle$ for any $h \in H$.

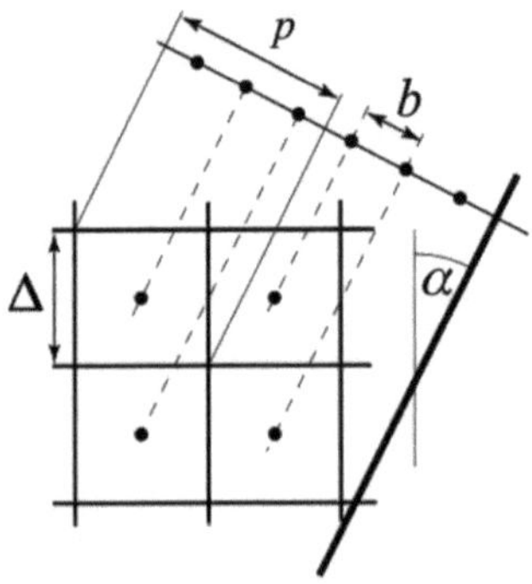

Fig. 8.8 Idealized setup for the determination of the MTF of a digital detector. Four-cell patch of the detector matrix and the edge line (*bold*) tilted at an angle α relative to the matrix. Centers of detector cells are projected onto the edge profile

$$\sigma^2_{\mathbf{P}q} \le \sigma^2_q \frac{C}{A}. \tag{8.61}$$

This approach can be applied, e.g., in the edge-based technique used for the determination of the pre-sampling *modulation transfer function* (MTF) of digital detectors. The setup used within the framework of this technique is depicted in Fig. 8.8, left.

The samples of the edge profile can be thought of to be the coefficients measured over the set $\{\mathrm{Sinc}(\Omega(x - bn))\}_n$, which is the frame of the space of band-limited functions. Usually, the edge profile is highly oversampled. As a consequence, the frame is redundant, and the redundancy factor A/C is very high. Let E be the smallest subspace of ℓ_2 such that $\mathrm{Im}\,\mathbf{U} \subset E$. Then, due to (8.61), the orthogonal projection onto E is supposed to reduce the noisy fraction of the edge profile significantly.

The linear independent frames constitute a special class of families called *Riesz basis*: the family $\{g_n\}$ is called Riesz basis of H if $H = \overline{\mathrm{span}\{g_n\}}$ and there exist constants $A > 0$ and $B < \infty$ such that

$$A \sum_n |c_n|^2 \le \left\| \sum_n c_n g_n \right\|^2 \le B \sum_n |c_n|^2 \tag{8.62}$$

for any $c \in \ell_2$. Note that the equivalency $\sum_n c_n g_n = 0 \;\Leftrightarrow\; c = 0$ that follows from (8.62) implies that elements of the Riesz basis are linearly independent.

For the Riesz basis constituted by integer translates of the function $g \in L_2(\mathbb{R})$, the condition (8.62) can be written as

$$A \le \sum_n |\hat{g}(\omega + 2\pi n)|^2 \le B, \quad \omega \in \mathbb{R}. \tag{8.63}$$

Since

$$\langle g, g(\cdot - k)\rangle = \frac{1}{2\pi}\int_{-\infty}^{\infty} |\hat{g}(\omega)|^2 e^{i\omega k} d\omega = \frac{1}{2\pi}\int_0^{2\pi} \sum_n |\hat{g}(\omega + 2\pi n)|^2 e^{i\omega k} d\omega, \quad (8.64)$$

one concludes that translates of g are orthonormal if and only if

$$\sum_n |\hat{g}(\omega + 2\pi n)|^2 \equiv 1. \quad (8.65)$$

In a similar way, it is verified that $\{g(\cdot - n)\}$ and $\{\tilde{g}(\cdot - n)\}$ are biorthogonal, i.e., $\langle g(\cdot - n), \tilde{g}(\cdot - k)\rangle = \delta_{n,k}$, if and only if

$$\sum_n \hat{g}(\omega + 2\pi n)\hat{\tilde{g}}(\omega + 2\pi n)^* \equiv 1. \quad (8.66)$$

Returning to wavelets, we accent that most known wavelet bases in $L_2(\mathbb{R})$ have been constructed by means of so-called *multi-resolution analysis* (MRA) formulated first in [8]. The key role within this approach is played by the so-called *scaling functions*: function $\varphi \in L_2(\mathbb{R})$ is called a scaling function if for any fixed $j \in \mathbb{Z}$ the family $\{\varphi_{j,n}\}_n$,

$$\varphi_{j,n}(x) = 2^{-j/2}(2^{-j}x - n), \quad (8.67)$$

is the Riesz basis of the subspace $V_j \subset L_2(\mathbb{R})$, and

(i)
$$V_j \subset V_{j-1}.$$

(ii)
$$f(\cdot) \in V_j \Leftrightarrow f(2\cdot) \in V_{j-1}.$$

(iii)
$$\overline{\bigcup_{j\in\mathbb{Z}} V_j} = L_2(\mathbb{R}), \ \bigcap_{j\in\mathbb{Z}} V_j = \{0\}.$$

Three necessary conditions for φ to be a scaling function are:

(1) The Fourier transform $\hat{\varphi}$ must satisfy (8.63).
(2) It satisfies the *two-scale relation*

$$\varphi = \sum_n h_n \varphi_{-1,n}. \quad (8.68)$$

(3) $$\hat{\varphi}(0) \neq 0.$$

In particular from (3), it follows that φ is a smoothing function. Therefore, V_j is called approximation, and the ladder $\ldots \subset V_{j+1} \subset V_j \subset V_{j-1} \subset \ldots$ as a multi-resolution approximation of $L_2(\mathbb{R})$.

Suppose that the Fourier transform of the scaling function φ satisfies (8.65), i.e., its integer translates are orthonormal. Let W_0 be the orthogonal compliment of V_0 in V_{-1}, that is, $V_{-1} = V_0 \oplus W_0$. Then there exists $\psi \in L_2(\mathbb{R})$, the integer translates of which constitute a basis of W_0. Moreover, ψ is necessarily a wavelet, the Fourier transform of which satisfies (8.65). In order to show this, let us rewrite the two-scale relation (8.68) in the Fourier domain:

$$\hat{\varphi}(\omega) = 2^{-1/2}\hat{h}(\omega/2)\hat{\varphi}(\omega/2), \tag{8.69}$$

where

$$\hat{h}(\omega) = \sum_n h_n \mathrm{e}^{-in\omega} \tag{8.70}$$

is 2π periodic. The orthonormality of $\varphi(\cdot - n)$ implies the condition

$$\left|\hat{h}(\omega)\right|^2 + \left|\hat{h}(\omega + \pi)\right|^2 \equiv 2, \tag{8.71}$$

which can be checked using (8.65). For any $f \in W_0$, the inclusion $f \in V_{-1}$ takes place, and consequently, there exists $a \in \ell_2$, so that $f = \sum_n a_n \varphi_{-1,n}$, or in the Fourier domain

$$\hat{f}(\omega) = 2^{-1/2}\hat{a}(\omega/2)\hat{\varphi}(\omega/2), \tag{8.72}$$

where $\hat{a}(\omega) = \sum_n a_n \mathrm{e}^{-i\omega n}$ is 2π periodic. The condition $W_0 \bot V_0$ implies that $\langle f, \varphi(\cdot - k)\rangle = 0$ for any $k \in \mathbb{Z}$, which in turn leads to the identity

$$\hat{a}(\omega)\hat{h}(\omega)^* + \hat{a}(\omega + \pi)\hat{h}(\omega + \pi)^* \equiv 0. \tag{8.73}$$

The possible solution of (8.73) is

$$\hat{a}(\omega) = \hat{v}(2\omega)\hat{h}(\omega + \pi)^*, \tag{8.74}$$

where $\hat{v}$ is any 2π-periodic function. Substitution of (8.74) into (8.72) yields

$$\hat{f}(\omega) = 2^{-1/2}\hat{v}(\omega)\hat{h}(\omega/2 + \pi)^*\hat{\varphi}(\omega/2), \tag{8.75}$$

or equivalently

$$f(x) = \sum_n v_n \psi(x-n), \tag{8.76}$$

where ψ is the function, the Fourier transform of which is

$$\hat{\psi}(\omega) = 2^{-1/2}\hat{h}(\omega/2+\pi)^*\hat{\varphi}(\omega/2). \tag{8.77}$$

Hence, we have shown that there exists ψ, translates of which constitute the basis in W_0. In order to see that ψ is a wavelet, note first that from (8.69), it follows that $\hat{h}(0) = \sqrt{2}$. Together with (8.71), this implies $\hat{h}(\pi) = 0$. Using this in (8.77) and taking into account that $\hat{\varphi}(0) \neq 0$ yields $\hat{\psi}(0) = 0$. That is, ψ is a wavelet (see condition (8.29)). The orthonormality of its translates can be checked by means of (8.65). Substituting (8.77) into (8.65), we obtain that

$$\sum_n |\hat{\psi}(\omega+2\pi n)|^2 = 2^{-1}\sum_n |\hat{h}(\omega/2+\pi n+\pi)^*\hat{\varphi}(\omega/2+\pi n)|^2. \tag{8.78}$$

Representing the sum on the right-hand side of (8.78) by two sums over odd and even indices, using (8.71) and accounting for the 2π periodicity of $\hat{h}$, we obtain

$$\sum_k |\hat{\psi}(\omega+\pi n)|^2 \equiv 1, \tag{8.79}$$

that is, $\{\psi(\cdot - n)\}_n$ is an orthonormal basis of W_0. This basis is not unique. Any λ defined by

$$\hat{\lambda}(\omega) = \hat{\alpha}(\omega)\hat{\psi}(\omega), \tag{8.80}$$

where $\hat{\alpha}$ is 2π periodic and $|\hat{\alpha}(\omega)| \equiv 1$, is a wavelet whose integer translates constitute the orthonormal basis in W_0. Indeed, $\hat{\lambda}(0) = 0$; moreover,

$$\sum_n |\hat{\lambda}(\omega+2\pi n)|^2 = |\hat{\alpha}(\omega)|^2 \sum_n |\hat{\psi}(\omega+2\pi n)|^2 \equiv 1. \tag{8.81}$$

In fact, any two functions, integer translates of which constitute orthogonal bases in W_0, relate to each other via (8.80) (see Chap. 8 of [6]). Therefore,

$$\hat{\psi}(\omega) = 2^{-1/2}\hat{\alpha}(\omega)\hat{h}(\omega/2+\pi)^*\hat{\varphi}(\omega/2), \tag{8.82}$$

where $\hat{\alpha}$ is 2π periodic and $|\hat{\alpha}(\omega)| \equiv 1$, characterizes all possible wavelets whose translates constitute the orthonormal basis in W_0. The choice of $\hat{\alpha}$ substantiates the wavelet. Usually, one sets $\hat{\alpha}(\omega) = \mathrm{e}^{-\mathrm{i}\omega/2}$. Then

$$\hat{\psi}(2\omega) = 2^{-1/2}\hat{g}(\omega)\hat{\varphi}(\omega), \tag{8.83}$$

where

$$\hat{g}(\omega) = \hat{h}(\omega + \pi)^* \mathrm{e}^{-\mathrm{i}\omega} \tag{8.84}$$

is the transfer function of coefficients of the decomposition of ψ in the basis of V_{-1}, that is,

$$\psi = \sum_n g_n \varphi_{-1,n}, \tag{8.85}$$

with

$$g_n = (-1)^n h_{1-n}. \tag{8.86}$$

From (i) and (ii) of MRA, it follows that the decomposition

$$V_{j-1} = V_j \oplus W_j \tag{8.87}$$

is valid for any j. Together with (iii), this yields

$$L_2(\mathbb{R}) = \bigoplus_{j\in\mathbb{Z}} W_j \tag{8.88}$$

Since $\{\psi_{j,k}\}_k$ with

$$\psi_{j,k} = 2^{-j/2}\psi(2^j x - k) \tag{8.89}$$

is the orthonormal basis of W_j, the whole family $\{\psi_{j,k}\}_{j,k}$ is the orthonormal basis of $L_2(\mathbb{R})$.

In order to construct the orthonormal wavelet basis in $L_2(\mathbb{R}^2)$, we first notice that the family $\{\varphi(\cdot - l)\varphi(\cdot - m)\}_{l,m}$ is the orthonormal basis of $V_0^2 = V_0 \otimes V_0$. It is also easy to see that for any j, the family $\{2^{-j}\varphi(2^{-j}\cdot - n)\varphi(2^{-j}\cdot - k)\}_{n,k}$ is the orthonormal basis of $V_j^2 = V_j \otimes V_j$ and $V_j^2 \subset V_{j-1}^2$. Using the fact that the set of all functions of view $f(x)g(y)$ is dense in $L_2(\mathbb{R}^2)$, we conclude that subspaces V_j^2 constitute multi-resolution approximation of $L_2(\mathbb{R}^2)$ in the sense (i)–(iii). Then, representing V_{j-1}^2 by the orthogonal sum

$$V_{j-1}^2 = V_j^2 \oplus W_j^2 \tag{8.90}$$

and recalling that $V_{j-1} = V_j \oplus W_j$, we obtain

$$(V_j \otimes V_j) \oplus (V_j \otimes W_j \oplus W_j \otimes V_j \oplus W_j \otimes W_j) = (V_j \otimes V_j) \oplus W_j^2, \tag{8.91}$$

which yields

$$W_j^2 = (V_j \otimes W_j) \oplus (W_j \otimes V_j) \oplus (W_j \otimes W_j). \tag{8.92}$$

Taking into account that $L_2(\mathbb{R}^2) = \bigoplus_j W_j^2$, we conclude that the family $\{\psi_{j,k}^1, \psi_{j,k}^2, \psi_{j,k}^3\}_{j\in\mathbb{Z},k\in\mathbb{Z}^2}$, where $\psi_{j,k}^i = 2^{-j}\psi^i(2^{-j}x - k_1, 2^{-j}y - k_2)$, and

$$\psi^1(x,y) = \varphi(x)\psi(y), \tag{8.93}$$

$$\psi^2(x,y) = \psi(x)\varphi(y), \tag{8.94}$$

$$\psi^3(x,y) = \psi(x)\psi(y), \tag{8.95}$$

constitutes the orthonormal basis in $L_2(\mathbb{R}^2)$.

MRA allows to fulfill the wavelet transformation of f by means of the cascade filter bank algorithm. We describe this algorithm for 1D. The extension to 2D is straightforward.

Rescaling (8.68) and (8.85) to $\varphi_{j,k} \sum_n h_{n-2k}\varphi_{j-1,n}$ and to $\psi_{j,k} \sum_n g_{n-2k}\varphi_{j-1,n}$, respectively, and denoting $f_{j,k} = \langle f, \varphi_{j,k}\rangle, d_{j,k} = \langle f, \psi_{j,k}\rangle$, one obtains the filter bank decomposition at step j:

$$\begin{aligned} f_{j,k} &= \sum_n f_{j-1,n}h_{n-2k}, \\ d_{j,k} &= \sum_n f_{j-1,n}g_{n-2k}. \end{aligned} \tag{8.96}$$

On the other hand, denoting with $\mathbf{P}A$ the operator of the orthogonal projection onto the subspace A, we can write

$$\mathbf{P}V_{j-1} = \mathbf{P}V_j + \mathbf{P}W_j. \tag{8.97}$$

Therefore, for any $f \in L_2(\mathbb{R})$,

$$\sum_n \langle f, \varphi_{j-1,n}\rangle\varphi_{j-1,n} = \sum_n \langle f, \varphi_{j,n}\rangle\varphi_{j,n} + \sum_n \langle f, \psi_{j-1,n}\rangle\psi_{j,n}. \tag{8.98}$$

Building the inner product of both sides of (8.98) with $\varphi_{j-1,k}$ yields the reconstruction formula

$$f_{j-1,k} = \sum_n f_{j,n} h_{k-2n} + \sum_n d_{j,n} g_{k-2n} \tag{8.99}$$

which is fulfilled recursively down to $j = 0$.

Let us consider an example of constructing orthonormal bases with box splines $\mathcal{B}_m$. Per definition, $\mathcal{B}_m$ is the $m - 1$ times convolution of $\mathcal{B}_1$ with itself, and

$$\mathcal{B}_1(x) = \begin{cases} 1 \text{if } 0 \leq \mathrm{x} < 1, \\ 0 \text{ otherwise.} \end{cases} \tag{8.100}$$

For $\mathcal{B}_m$, it is known that its integer translates constitute Riesz basis of $V_0^m = \overline{L_2(\mathbb{R}) \cap S_m^0}$ with the Riesz-bounds A_m and B_m

$$A_m = 2\pi^{-2m}(2^{2m} - 1)\sum_{k=1}^{\infty} k^{-2m},\ B_m = 1 \tag{8.101}$$

(see, e.g., Chap. 4 of [9]). The two-scale relation for $\mathrm{B_m}$ is known to be

$$\mathcal{B}_m(x) = \sum_{n=0}^{m} 2^{-m+1/2} \binom{m}{n} \mathcal{B}_m(2x - n). \tag{8.102}$$

Since $\hat{\mathcal{B}}_m(\omega) = (\hat{\mathcal{B}}_1(\omega))^{m-1}$ and

$$\hat{\mathcal{B}}_1(\omega) = \frac{\sin(\omega/2)}{\omega/2} \mathrm{e}^{-\mathrm{i}\omega/2}, \tag{8.103}$$

we obtain that $\hat{\mathcal{B}}_m(0) = 1$. That is, $\mathcal{B}_m$ is a scaling function that generates a multi-resolution approximation $\ldots \subset V_j^m \subset V_{j-1}^m \subset \ldots$, where $V_j^m = \overline{L_2(\mathbb{R}) \cap S_m^j}$.

By $m = 1$, the coefficients of the two-scale relation (8.102) are $h_0 = h_1 = 2^{-1/2}$. The wavelet associated with $\mathcal{B}_1$, constructed with the help of (8.85) and (8.86), is

$$\psi(x) = \mathcal{B}_2(2x) - \mathcal{B}_1(2x - 1) = \begin{cases} 1 & \text{if } 0 \leq \mathrm{x} \leq 1/2, \\ -1 & \text{if } 1/2 \leq \mathrm{x} \leq 1, \\ 0, & \text{otherwise.} \end{cases} \tag{8.104}$$

This is the well-known Haar wavelet. Since Riesz-bounds $A_1 = B_1 = 1$, the translates $\mathcal{B}_1(\cdot - n)$ are orthonormal (see above (8.63) and (8.65), and consequently so are the translates $\psi(\cdot - n)$. The whole family $\{2^{-j/2}\psi(2^{-j} \cdot - n)\}_{j,n}$ is the orthonormal basis of $L_2(\mathbb{R})$.

As it follows from (8.101), the lower Riesz-bound A_m decays with m. Therefore, $A_m < 1$ for any $m > 1$, and as a consequence, the integer translates of $\mathcal{B}_m$ are not orthonormal. Applying to $\mathcal{B}_m$ the orthogonalization trick

$$\hat{\mathcal{B}}^{\#}_m(\omega) = \hat{\mathcal{B}}_m(\omega)\left(\sum_n |\hat{\mathcal{B}}_m(\omega + 2\pi n)|^2\right)^{-1/2}, \quad \omega \in R, \tag{8.105}$$

one obtains functions $\mathcal{B}^{\#}_m$ which satisfy (8.65), and as a consequence, their integer translates are orthonormal. However, these functions are no more compactly supported, and neither are the associated wavelets called Battle-Lemarié wavelets.

In practical applications, numerically advantageous compactly supported wavelets are preferable. In applications where the perceptual assessment of the signal is important, e.g., medical imaging, the symmetry of wavelets is another important feature of preference: applying the asymmetric wavelets can yield asymmetric errors,[4] and it is known that the human visual system is less tolerant to asymmetric errors than to symmetric ones. Additionally, symmetric or antisymmetric wavelets are more efficient while treating boundaries of the compactly supported signal, such as an image. For such applications, the basis of $L_2(T)$, where $T \subset \mathbb{R}^n$ is supposed to be the support of the signal, has to be constructed. Normally this is done by modifying wavelets of the basis of $L_2(\mathbb{R}^n)$: wavelets, the support of which is entirely inside T are not changed; the wavelets, the support of which is entirely outside T are skipped; and those whose support overlaps the boundaries of T are modified. There are different ways to modify the boundary wavelets (see, e.g., Chap. 7 of [7]). For symmetric or antisymmetric wavelets, boundary wavelets are simply folded back away from boundaries. The decomposition in the so-obtained basis is equivalent to the decomposition of the signal that is symmetrically extended beyond T. An example of such extension for 1D signal supported on $T = [a, b]$ is given in Fig. 8.9.

It is well known (see, e.g., [6]) that constructing the orthonormal basis from compactly supported wavelet is possible only if the wavelet is asymmetric. The exception is the discontinuous Haar wavelet. For any continuous compactly supported wavelet, the symmetry and orthonormality are inconsistent. But compactly supported symmetric or antisymmetric wavelets can be used to construct biorthogonal bases of $L_2(\mathbb{R})$.

For the following we accept (without proof) several important facts about compactly supported scaling functions. Let h be the filter the transfer function of which is the trigonometric polynomial

$$\hat{h}(\omega) = \sum_{n=N_1}^{N_2} h_n \mathrm{e}^{-\mathrm{i}\omega n}. \tag{8.106}$$

In other words, h is a finite impulse response (FIR) filter, i.e., the filter finitely many taps of which are not zero. Let in addition

$$\hat{h}(0) = \sqrt{2} \quad \text{and} \quad \hat{h}(\pi) = 0. \tag{8.107}$$

[4] That is, errors that are described with asymmetric density distribution.

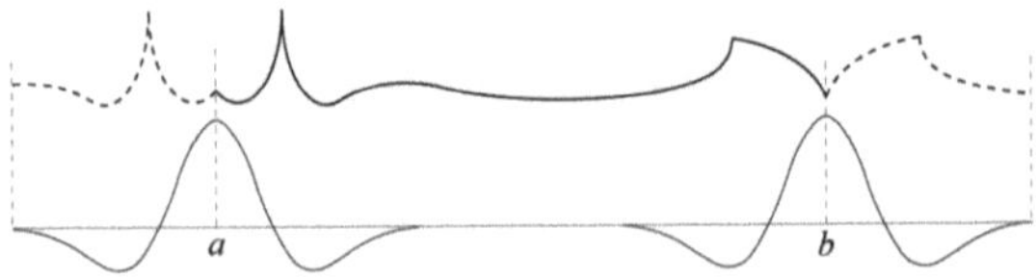

Fig. 8.9 To the boundary problem of the decomposition of the signal supported on the limited interval. The signal (*bold*) and its extension (*dashed*); two boundary wavelets located at most extreme positions

Then φ, the Fourier transform of which is

$$\hat{\varphi}(\omega) = \prod_{j=1}^{\infty} \frac{\hat{h}(2^{-j}\omega)}{\sqrt{2}}, \tag{8.108}$$

is a compactly supported square integrable scaling function. The function defined by (8.108) will be referred to as the one related to h. A pair of FIR low-pass filters h and $\tilde{h}$, both satisfying (8.107), are called dual if the families $\{\varphi(\cdot - n)\}_n$ and $\{\tilde{\varphi}(\cdot - n)\}_n$, where φ and $\tilde{\varphi}$ are related scaling functions, are biorthogonal. The necessary condition for h and $\tilde{h}$ to be dual is

$$\hat{h}(\omega)^{\bullet}\hat{\tilde{h}}(\omega) + \hat{h}(\omega+\pi)^{\bullet}\hat{\tilde{h}}(\omega+\pi) \equiv 2 \tag{8.109}$$

(for sufficient conditions, see [10]).

The multi-resolution analysis can be reformulated in terms of biorthogonal bases. Let h and $\tilde{h}$ be dual filters, and let the related scaling functions generate multi-scale approximations of $L_2(\mathbb{R})$

$$\ldots \subset V_j \subset V_{j-1} \subset \ldots \tag{8.110}$$

and

$$\ldots \subset \tilde{V}_j \subset \tilde{V}_{j-1} \subset \ldots \tag{8.111}$$

respectively.

Denote with W_j the subspace that completes the subspace V_j in V_{j-1} and with $\tilde{W}_j$ the complement subspace of $\tilde{V}_j$. Apparently $W_0 \perp \tilde{V}_0$ and $\tilde{W}_0 \perp V_0$. In the similar way as we did in the orthonormal case, but using (8.66) instead of (8.65), it is possible to show that there exist biorthogonal Riesz bases of W_0 and $\tilde{W}_0$ that are constituted with integer translates of wavelets ψ and $\tilde{\psi}$, the Fourier transform of which are

$$\hat{\psi}(\omega) = 2^{-1/2}\hat{\alpha}(\omega)\hat{\tilde{h}}(\omega/2+\pi)^{*}\hat{\varphi}(\omega/2), \tag{8.112}$$

$$\hat{\tilde{\psi}}(\omega) = 2^{-1/2}\hat{\beta}(\omega)\hat{h}(\omega/2+\pi)^{*}\hat{\tilde{\varphi}}(\omega/2), \tag{8.113}$$

with $\hat{\alpha}$ and $\hat{\beta}$ being 2π periodic, $|\hat{\alpha}(\omega)| \equiv 1$, $|\hat{\beta}(\omega)| \equiv 1$, and $\hat{\alpha}(\omega)\hat{\beta}(\omega)^* \equiv 1$. The families $\{2^{-j/2}\psi(2^{-j}\cdot - n)\}_{j,n}$ and $\{2^{-j/2}\tilde{\psi}(2^{-j}\cdot - n)\}_{j,n}$ are biorthogonal Riesz bases of $L_2(\mathbb{R})$.

For $\hat{\alpha}(\omega) = \mathrm{e}^{-\mathrm{i}\omega/2}$, the equalities (8.112) and (8.113) in the space domain are

$$\psi = \sum_n g_n \varphi_{-1,n} \; and \; \tilde{\psi} = \sum_n \tilde{g}_n \tilde{\varphi}_{-1,n} \tag{8.114}$$

with

$$\begin{aligned} g_n &= \langle \psi, \tilde{\varphi}_{-1,n} \rangle = (-1)^{n+1} \tilde{h}_{1-n} \quad \text{and} \\ \tilde{g}_n &= \langle \tilde{\psi}, \varphi_{-1,n} \rangle = (-1)^{n+1} h_{1-n}. \end{aligned} \tag{8.115}$$

The extension to the 2D case is made in a similar way as we did it in the orthonormal case. One defines three wavelets ψ^1, ψ^2, and ψ^3 exactly as in (8.93)–(8.95), which in this case generate the Riesz basis of $L_2(\mathbb{R}^2)$. The wavelets $\tilde{\psi}^1(x,y) = \tilde{\varphi}(x)\tilde{\psi}(y)$, $\tilde{\psi}^2(x,y) = \tilde{\psi}(x)\tilde{\varphi}(y)$, and $\tilde{\psi}^3(x,y) = \tilde{\psi}(x)\tilde{\psi}(y)$ generate the dual Riesz basis of $L_2(\mathbb{R}^2)$.

In the same way as in the orthonormal case, the two-scale relation on the one hand and the relations (8.114) on the other imply the fast filter bank algorithm of the wavelet transform of f:

$$\langle f, \varphi_{j,k} \rangle = \sum_m \langle f, \varphi_{j-1,m} \rangle h_{m-2k}, \tag{8.116}$$

$$\langle f, \psi_{j,k} \rangle = \sum_m \langle f, \psi_{j-1,m} \rangle g_{m-2k}. \tag{8.117}$$

In order to obtain the reconstruction formula, let us introduce the operator

$$\mathbf{P}_V f = \sum_k \langle f, \lambda_k \rangle \tilde{\lambda}_k, \tag{8.118}$$

where $\{\lambda_k\}_k$, $\{\tilde{\lambda}_k\}_k$ are biorthonormal Riesz bases of the subspace V and $\tilde{V}$, respectively. The operator $\mathbf{P}_V$ is an orthoprojector. Since the families $\{\varphi_{j,n}, \psi_{j,n}\}_n$ and $\{\tilde{\varphi}_{j,n}, \tilde{\psi}_{j,n}\}_n$ are biorthogonal Riesz bases of V_{j-1} and $\tilde{V}_{j-1}$, respectively, we obtain that $\mathbf{P}V_{j-1} = \mathbf{P}V_j + \mathbf{P}W_j$, that is, for any $f \in L_2(\mathbb{R})$, the identity

$$\sum_n \langle f, \varphi_{j-1,n} \rangle \tilde{\varphi}_{j-1,n} = \sum_n \langle f, \varphi_{j,n} \rangle \tilde{\varphi}_{j,n} + \sum_n \langle f, \psi_{j,n} \rangle \tilde{\psi}_{j,n} \tag{8.119}$$

is valid. The inner product of (8.119) with $\varphi_{j-1,k}$ yields the reconstruction formula

$$\langle f, \varphi_{j-1,k} \rangle = \sum_n \langle f, \varphi_{j,n} \rangle \tilde{h}_{k-2n} + \sum_n \langle f, \psi_{j,n} \rangle \tilde{g}_{k-2n}. \tag{8.120}$$

We conclude our review by considering the redundant dyadic wavelet transform that is defined by

$$Wf_j(u) = \langle f, \psi_j(\cdot - u) \rangle, \; u \in \mathbb{R}, \; j \in \mathbb{Z}, \tag{8.121}$$

where $\psi_j(x) = 2^{-j/2}\psi(2^{-j}x)$. In a similar way as we did before, let us introduce the self-adjoint operator $\mathbf{H} : L_2(\mathbb{R}) \to L_2(\mathbb{R})$, defined by

$$\mathbf{H}f(t) = \sum_j 2^{-j} \int Wf_j(u)\psi_j(t-u)\mathrm{d}u. \tag{8.122}$$

If there exist constants $0 < A \leq B < \infty$ such that the bounding condition

$$A\|f\|^2 \leq \langle \mathbf{H}f, f \rangle \leq B\|f\|^2 \tag{8.123}$$

holds for all $f \in L_2(\mathbb{R})$, then the operator $\mathbf{H}$ is invertible with the inverse $\mathbf{H}^{-1}$ bounded by B^{-1} and A^{-1}. The sufficient condition for $\mathbf{H}$ to satisfy (8.123) is

$$A \leq \sum_j \left|\hat{\psi}(2^j\omega)\right|^2 \leq B. \tag{8.124}$$

In order to see this, the inequality (8.124) has to be multiplied by $|\hat{f}(\omega)|^2$ and integrated over ω. Then, using the representation

$$\widehat{Wf}_j(\omega) = 2^{j/2}\hat{\psi}(2^j\omega)^*\hat{f}(\omega) \tag{8.125}$$

and taking into account the Parseval's identity, one obtains

$$A\|f\|_{L_2}^2 \leq \sum_j 2^{-j}\|Wf_j\|_{L_2}^2 \leq B\|f\|_{L_2}^2, \tag{8.126}$$

which is another equivalent notation of (8.123).

Applying $\mathbf{H}^{-1}$ to both sides of (8.122) yields

$$f(t) = \sum_j 2^{-j} \int Wf_j(u)\tilde{\psi}_j(t-u)\mathrm{d}u, \tag{8.127}$$

where $\tilde{\psi} = \mathbf{H}^{-1}\psi$ is called a dual wavelet. In order to determine $\mathbf{H}^{-1}$, one has to notice that

$$\widehat{\mathbf{H}f}(\omega) = \hat{f}(\omega) \sum_j |\hat{\psi}(2^j\omega)|^2, \tag{8.128}$$

that is, in the Fourier domain, the operator $\mathbf{H}$ is associated with the operator $\hat{\mathbf{H}}$, which is simply the multiplicator $\sum_j |\hat{\psi}(2^j\omega)|^2$. Therefore,

$$\hat{\mathbf{H}}^{-1} = \frac{1}{\sum_j |\hat{\psi}(2^j\omega)|^2}, \tag{8.129}$$

and consequently,

$$\hat{\tilde{\psi}}(\omega) = \hat{\mathbf{H}}^{-1}\hat{\psi}(\omega) = \frac{\hat{\psi}(\omega)}{\sum_j |\hat{\psi}(2^j\omega)|^2} \tag{8.130}$$

which makes sense due to (8.124). In fact, any $\tilde{\psi}$ satisfying

$$\sum_j \hat{\tilde{\psi}}(2^j\omega)\hat{\psi}(2^j\omega)^* = 1, \omega \in \mathbb{R}\backslash\{0\}, \tag{8.131}$$

satisfies also (8.127). This follows from the fact that in the Fourier domain, the relation (8.127) is represented by

$$\hat{f}(\omega) = \hat{f}(\omega) \sum_j \hat{\tilde{\psi}}(2^j\omega)\hat{\psi}(2^j\omega)^*. \tag{8.132}$$

Let h and g be a pair of FIR filters, and let $\hat{h}(0) = \sqrt{2}$. Let φ be the scaling function that relates to h via (8.108). Then, apparently,

$$\hat{\varphi}(2\omega) = 2^{-1/2}\hat{h}(\omega)\hat{\varphi}(\omega). \tag{8.133}$$

The Fourier transform of the corresponding wavelet is defined by

$$\hat{\psi}(2\omega) = 2^{-1/2}\hat{g}(\omega)\hat{\varphi}(\omega). \tag{8.134}$$

Let $\tilde{h}$ and $\tilde{g}$ be, in turn, another pair of FIR filters, and $\tilde{\varphi}$ and $\tilde{\psi}$ are the related scaling function and the corresponding wavelet. Since both φ and $\tilde{\varphi}$ are compactly supported, they are from $L_1(\mathbb{R}) \cap L_2(\mathbb{R})$. Therefore, their Fourier transforms are continuous. This fact can be used to show that the condition

$$\hat{\tilde{h}}(\omega)\hat{h}(\omega)^* + \hat{\tilde{g}}(\omega)\hat{g}(\omega)^* = 2 \tag{8.135}$$

is sufficient for ψ and $\tilde{\psi}$ to satisfy (8.131).

Hence, given h and g, one has a certain freedom in choosing filters $\tilde{h}$ and $\tilde{g}$. For example, one can set $\tilde{h} = h$. Then from (8.135), we obtain that

$$\hat{\tilde{g}}(\omega) = \frac{2 - |\hat{h}(\omega)|^2}{\hat{g}(\omega)^*}. \tag{8.136}$$

The relation (8.136) imposes certain restrictions on g. Namely, zeros of $\hat{g}$ must coincide with zeros of $2 - |\hat{h}(\omega)|^2$ and be of the corresponding order. In particular at $\omega = 0$, there must be one zero of $\hat{g}$. Since the number of zeros of $\hat{g}$ at $\omega = 0$ coincides with the number of zeros of $\hat{\psi}$ at $\omega = 0$, it follows that the corresponding wavelet ψ must have one vanishing moment.

A described strategy was applied in [11] for the construction of the so-called *Mallat-wavelet*. Redundant wavelet family generated with this wavelet has been proven to be efficient for the reconstruction of the image from the wavelet coefficients measured over multi-scale edges. The starting point while constructing the wavelet in [11] was to set

$$\varphi(x) = \mathcal{B}_3(x+1), \tag{8.137}$$

where $\mathcal{B}_3$ is the box spline of degree 2. The Fourier transform of φ is

$$\hat{\varphi}(\omega) = \left(\frac{\sin(\omega/2)}{\omega/2}\right)^3 \mathrm{e}^{-\mathrm{i}\omega/2}. \tag{8.138}$$

We know that $\mathcal{B}_3$ satisfies the two-scale relation (see (8.102)), and so consequently does φ. That is, there exists a filter h such that $\varphi = \sum_n h_n \varphi_{-1,n}$. The transfer function of h is yielded from (8.133):

$$\hat{h}(\omega) = \sqrt{2}\frac{\hat{\varphi}(2\omega)}{\hat{\varphi}(\omega)} = \sqrt{2}\cos^3\frac{\omega}{2}\mathrm{e}^{-\mathrm{i}\omega/2}. \tag{8.139}$$

Let us turn our attention to the fact, that the Mallat-wavelet was constructed using the additional requirement to be efficient for the detection of edges. From (8.134) and the fact that both g and h are FIR filter, it follows that ψ is compactly supported. Since ψ defined by (8.136) must have one vanishing moment, then due to (8.37), there must exist compactly supported smoothing function θ such that

$$\psi = -\theta'. \tag{8.140}$$

Due to (8.38), we obtain then that $Wf_j(u) = 2^j \frac{\mathrm{d}}{\mathrm{d}t} f * \bar{\theta}_j(t)$, so that local maxima of $|Wf_j(u)|$ are exactly local maxima of $|\mathrm{d}/\mathrm{d}x(f * \bar{\theta}_j)|$. The latter are edge points within

Canny's approach of an edge detection [12]. Hence, the desired wavelet is determined by the choice of θ. For example, one can set

$$\theta = \varphi. \tag{8.141}$$

However, this choice leads to a piecewise linear wavelet. Since the regularity of the wavelet is crucial for the quality of reconstruction, in [11], θ was set to

$$\theta(x) = 2^{-1}\mathcal{B}_4(2x+1). \tag{8.142}$$

The Fourier transform of the wavelet $\psi = -\theta'$, referred to as the Mallat-wavelet, is

$$\hat{\psi}(\omega) = -i\frac{\omega}{4}\left(\frac{\sin(\omega/4)}{\omega/4}\right)^4 \mathrm{e}^{-\mathrm{i}\omega/2}. \tag{8.143}$$

The transfer function of the corresponding filter g, obtained using (8.134) and (8.138), is

$$\hat{g}(\omega) = -i\sqrt{2}\sin\frac{\omega}{2}\mathrm{e}^{-\mathrm{i}\omega/2}. \tag{8.144}$$

Finally using (8.136), for the transfer function of the reconstruction filter $\tilde{g}$, one obtains

$$\hat{\tilde{g}}(\omega) = -i\sqrt{2}\sin\frac{\omega}{2}\left(\sum_{k=0}^{3}\cos^{2k}\frac{\omega}{2}\right)\mathrm{e}^{-\mathrm{i}\omega/2}. \tag{8.145}$$

Using pairs of filters h, g and $\tilde{h}$, $\tilde{g}$, a fast dyadic wavelet transform can be calculated with a filter bank algorithm called *algorithme à trous* [13]. The corresponding filter bank algorithm in 2D is

$$f_{j+1}(m,n) = f_j * \bar{h}_j\bar{h}_j(m,n), \tag{8.146}$$

$$d^1_{j+1}(m,n) = f_j * \bar{g}_j\delta(m,n), \tag{8.147}$$

$$d^2_{j+1}(m,n) = f_j * \delta\bar{g}_j(m,n), \tag{8.148}$$

where $\alpha\beta(m,n)$ is a separable two-dimensional filter, i.e., $\alpha\beta(m,n) = \alpha(m)\beta(n)$; $\delta(n)$ is a discrete Dirac; and α_j is the filter obtained from α by inserting $2^j - 1$ zeros (*trous*) between its samples. The reconstruction is made recursively by means of

$$f_j(m,n) = f_{j+1} * h_jh_j(m,n) + d^1_{j+1} * \tilde{g}_j\tilde{q}_j(m,n) + d^2_{j+1} * \tilde{q}_j\tilde{g}_j(m,n), \tag{8.149}$$

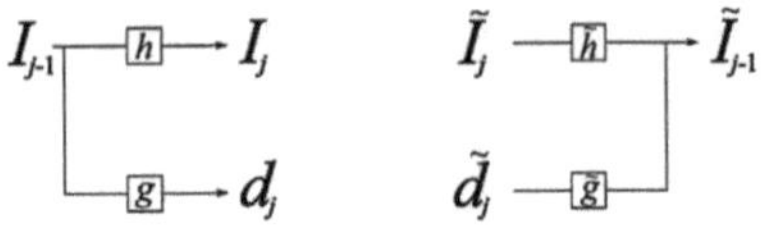

Fig. 8.10 Analysis (*left*) and synthesis (*right*) steps of the filter bank filtration

where $\tilde{g}$ is defined by (8.136), and for $\tilde{q}$, we have

$$\hat{\tilde{q}}(\omega) = \frac{1 + |\hat{h}(\omega)|^2}{2} \tag{8.150}$$

(see [11]).

We have described several of the most often applied wavelet decomposition-reconstruction frameworks. All of them can be implemented with fast filter bank algorithms, the common principal scheme of which is depicted in Fig. 8.10.

In the following, we give some heuristic considerations concerning the question of how these schemes can be used for the reduction of noise. As one can see in Fig. 8.10, during the analysis step, the coarse approximation I_{j-1} is further decomposed in the coarser approximation I_j and the details image d_j that contains the wavelet coefficients $\langle I, \psi_{j,k} \rangle$. Due to the localization properties of wavelets, one can change the wavelet coefficients within some region R of the image d_j, not bothering much about what effect this would have for those structures of the synthesized image $\tilde{I}_{j-1}$ which lie apart from R. If, for example, we want to reproduce the region R of the image I_j in the image $\tilde{I}_{j-1}$, we should suppress all details of d_j located in R, that is, the corresponding wavelet coefficients have to be set to zero. In order to suppress noisy details of the image I_{j-1}, we need to know which detail is due to noise.

There are pretty different scenarios to decide whether (or not) the given wavelet coefficient is due to noise. We mention an approach developed in [14]. The idea lying in the background of the method proposed there is to extract *relevant* details by means of comparative analysis of two or several images of the same object. It will be not quite far from the truth to say, that the comparative analysis is what the radiologist does while investigating the image. Evaluating the X-ray image, the observer automatically extracts features which he qualifies as *relevant*. Speaking figuratively, the skilled observer *correlates* a real image with an imaginary one that was created by his visual system. Motivated by the wish to facilitate the daily task of the radiologist, the authors have decided to apply the "correlation approach" in clinical practice. As soon as we have two images of the same object, we can define as relevant those details, which are present in both of them. The other details are qualified as noise and suppressed. In [14] the results of the method applied to quasi identical as well as to those which are slightly deformed relative to each other were reported. The technical aspect of the method is as follows. The both images, say A and B, are decomposed using the dyadic wavelet frame generated by the Mallat mother wavelet. The reason for the choice of this wavelet is the shift-invariance of the related wavelet transform. After the decomposition, details of

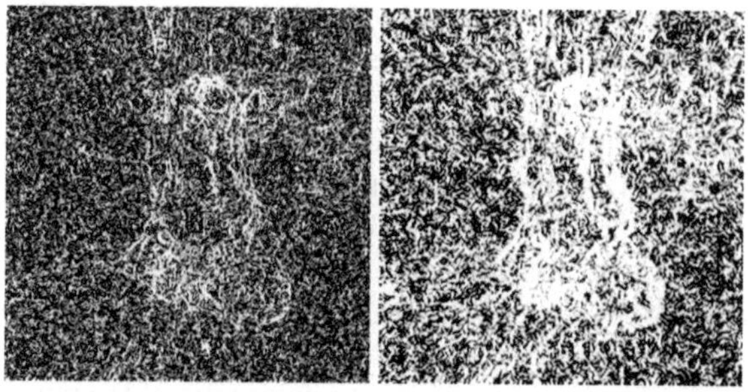

Fig. 8.11 Weighting masks appearing as a result of comparative analysis of details images

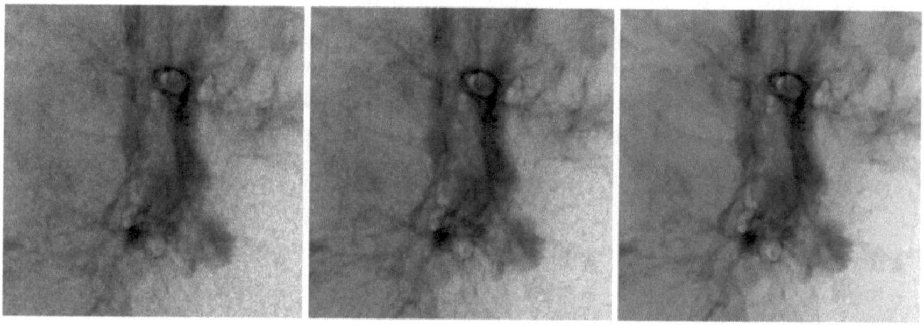

Fig. 8.12 Denoising of the lung phantom. *Left*: one of the two images of the same object. *Middle*: the sum of both images. *Right*: the image reconstructed from details weighted with masks shown in Fig. 8.11

image A and image B are correlated with each other at each resolution level. The result of the correlation is the *weighting matrix*, each element of which expresses a measure of similarity between the detail of image A and the detail of image B at the corresponding location. Denoting the measure of similarity with $w^j_{m,k}$, where j is the resolution level and (m,k) the location within the image, we have

$$0 \leq w^j_{m,k} \leq 1. \tag{8.151}$$

If $w^j_{m,k} = 1$, details of A and B at the level j and location (m,n) coincide; the case $w^j_{m,k} = 0$ means that the corresponding detail is due to noise; the intermediate cases should point at the portion of the noise at the location. Here we demonstrate the functionality of the method applying it to images of a lung phantom, the patch from the one of which is shown on the left of Fig. 8.4. Figure 8.11 shows weighting matrices of the first two resolution levels.

During the following step, elements of detail images are weighted by elements of corresponding weighting matrices. The final image is reconstructed from weighted coefficients. The result of the reconstruction is shown on the right-hand side of Fig. 8.12.

In spite of the seemingly disadvantage expressed by the necessity to have two or more images, the method provides a reduction of noise with minimal risk to lose useful information. In practice, this method could be applied to two images each acquired at half doses, for example, by means of detector based on CCD camera.

In this chapter, we described the mathematical foundations of image denoising and its practical implications. The basic principles are also valid for three-dimensional imaging and can in general be performed in all types of clinical diagnostic imaging. In nuclear medical imaging, the consequences of image denoising have to be studied in very much detail due to the fact that in nuclear medical imaging, the signal-to-noise ratio of the raw data is very poor to avoid high doses of radiation to the patient. This implies that unreflected use of denoising methods particularly also those based on compressing information can result in mainly reducing relevant image structures together with or even instead of the noise. Which kind of noise reduction can be used without deteriorating necessary image information has to be studied for each single application.

References

1. Zygmund A (1959) Trigonometric series, chapter X, vol 2. Cambridge University, Cambridge
2. Oppelt A (2005) Imaging systems for medical diagnostics: fundamentals, technical solutions, and applications for systems applying ionizing radiation, nuclear magnetic resonance, and ultrasound. Wiley-VCH Verlag GmbH, New York, 2/E
3. Wiener N (1949) Extrapolation, interpolation, and smoothing of stationary time series. Wiley, New York
4. Tomasi C, Manduchi R (1998) Bilateral filtering for gray and color images. In: ICCV'98: proceedings of the sixth international conference on computer vision. IEEE Computer Society, Washington, DC, p 839
5. Aurich V, Weule J (1995) Non-linear Gaussian filters performing edge preserving diffusion. In: Proceedings of the 7th DAGM-Symposium, Bielefeld, Springer, pp 538–545
6. Daubechies I (1992) Ten lectures on wavelets. SIAM, Philadelphia, PA
7. Mallat S (1999) A wavelet tour of signal processing. Academic, San Diego, CA
8. Mallat S (1989) A theory for multiresolution signal decomposition: the wavelet representation. IEEE Trans Pattern Anal Mach Intell 11(7):674–693
9. Chui CK (1992) Introduction to wavelets. Academic, London
10. Cohen A, Daubechies I, Feauveau JC (1992) Orthonormal bases of compactly supported wavelets. Commun Pure Appl Math 45:485–560
11. Mallat S, Zhong S (1992) Characterization of signals from multiscale edges. IEEE Trans Pattern Anal Mach Intell 4(7):70–732
12. Canny J (1986) A computational approach to edge detection. IEEE Trans Pattern Anal Mach Intell 36:961–1005
13. Holschneider M, Kronland-Martinet R, Morlet J, Tchamitchian P (1989) A real time algorithm for signal analysis with help of the wavelet transform. In: Combes JM, Grossmann A, Tchamitchian P (eds) Wavelets: time frequency methods and phase space. Springer, Berlin, pp 286–297
14. Tischenko O, Hoeschen C, Buhr E (2005) Reduction of anatomical noise in medical X-ray images. Radiat Prot Dosimetry 4(Nos ß3):69–74

Chapter 9
New Approaches to Quality Assurance

Hugo de las Heras Gala

9.1 Introduction

Any kind of technology needs some kind of quality assurance program. As an example, a bicycle needs to be checked regularly, and sometimes, one must have it thoroughly checked by professionals. Otherwise, a sudden failure of some part may lead to serious consequences for the rider or for anyone close by.

The World Health Organization defines a quality assurance or quality control (QA/QC) program as "an organized effort by the staff operating a facility to ensure that the diagnostic images produced are of sufficiently high quality, so that they consistently provide adequate diagnostic information at the lowest possible cost and with the least possible exposure of the patient to radiation" [1].

In fact, a slight deviation from the perfect performance of a medical imaging device would affect a large amount of people, since every device in a hospital is used on many patients within short periods of time. Deviations from the appropriate settings of a device can be due to faulty parts (e.g., a tube anode being too old) but most often to the staff handling the device. For instance, a thicker patient may require a higher value of the tube current to acquire a good image of the intestines, and without proper QA programs, the wrong setting could remain for subsequent patients. Since devices need to be adjusted for specific patients almost every day, QA checks must be strict, accurate, and regular. Standard documents for the quality assurance of single nuclear medicine scanners [2, 3] are available. However, at present, almost all nuclear medicine scanners are included in hybrid systems, as described in the previous chapters. These systems provide functional information from a pure nuclear medicine scan (PET or SPECT) together with anatomical

H. de las Heras Gala (✉)
US Food and Drug Administration, 10903 New Hampshire Avenue, 20993 Silver Spring, MD, USA

QUART GmbH, Kirchenweg 7, 85604 Zorneding, Germany
e-mail: Hugo.heras@daad-alumni.de

A. Giussani and C. Hoeschen (eds.), *Imaging in Nuclear Medicine*,
DOI 10.1007/978-3-642-31415-5_9, © Springer-Verlag Berlin Heidelberg 2013

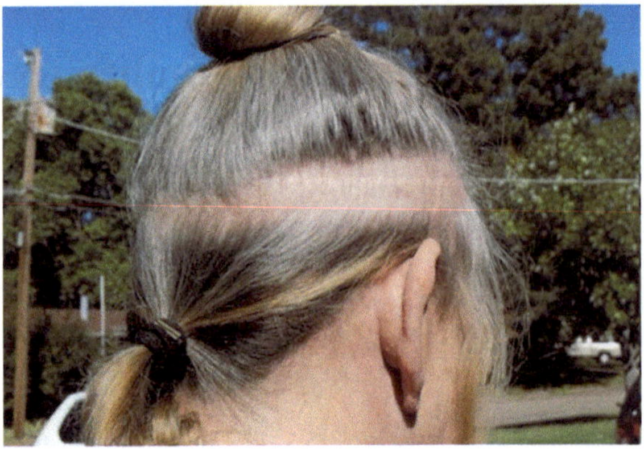

Fig. 9.1 The effect of X-rays 2 weeks after of a wrongly set CT perfusion scan. Photo from al.com

information from an X-ray (CT) or magnetic resonance (MR) scan. Therefore, in this chapter, we will also deal with quality assurance also for conventional X-ray radiology and occasionally for magnetic resonance.

Despite accidents with radiology devices are very seldom, the interest in QA for the particular case of computed tomography devices has increased dramatically since 2010 because of the effects of a big mistake that happened in a clinic in the USA [4, 5].

Apparently, a technologist intended to reduce the radiation dose to the patients by allowing a larger amount of noise in the images, but he did by mistake the opposite. A stripe of hair loss in the heads of the patients (Fig. 9.1) appeared after a couple of weeks, but it was difficult for the patients and their doctors to link those effects to their past CT examination. During almost a year, patients were scanned using wrong settings until the cases appeared in the news. This event has recently urged the need for consistent and frequent QA programs in all radiological facilities.

Any program for QA in medical imaging involves measurements of radiation dose and image quality.

9.1.1 The Risk of Radiation Dose

The current international approach to every kind of radiation is based on the ALARA principle (as low as reasonably achievable), which itself is based on the assumption that the risk of a certain amount of radiation is proportional to its magnitude. This assumption (called linear non-threshold model [6, 7]) seems reasonable, but it has been recently challenged by *hormesis* theories [8], which defend that a slight amount of radiation may be *beneficial*; that low dose may prepare your body's mechanisms against larger amounts of radiation (Fig. 9.2), just like a vaccination does against a virus. Other positions (called supralinear models) defend that low doses of radiation have actually a relatively *higher risk* than large doses.

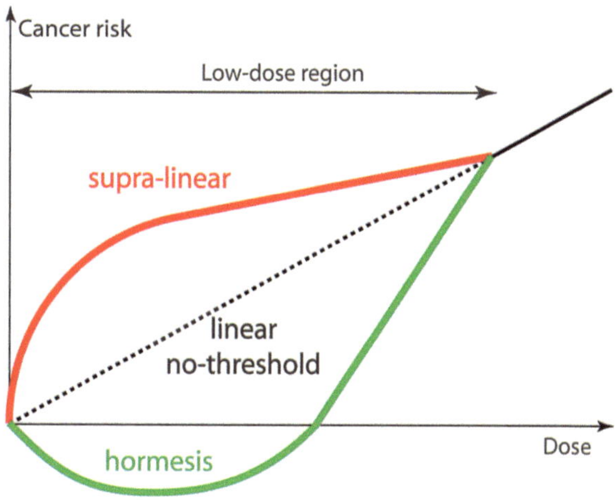

Fig. 9.2 Theories of cancer risk vs. dose

Experiments to check these theories are extremely difficult because the involved doses are in the range of the naturally received radiation. A group of individuals taking part in a control study, for example, would have to be isolated from every kind of radiation, which would isolate them from everything else at the same time, and therefore, it would be difficult to find a control group for them. Apart from that, individuals show different sensitivity to radiation doses depending on their genes. This individual susceptibility makes careful studies even more difficult.

For this reason, the author joins the majority of the scientific community assuming the LNT model and defending the ALARA principle. Accordingly, current trends for radiation dosimetry in QA programs are discussed in Sect. 9.2.

9.1.2 Image Quality

Traditional approaches to determine image quality (IQ) start to be obsolete, either because they rely on subjective tests or because they have been proven to miss features of the images. In particular, new technologies, such as noise reduction algorithms for CT, or mixed devices that combine transmission and emission tomography, such as PET-CT or PET-MR scanners, require new assessment tools and models and new phantoms to appropriately account for those new features. Some of these new tools will be presented in Sect. 9.3.

An alternative current trend for IQ evaluation, although suggested more than 30 years ago, is the use of mathematical or *model observers*. These tools can be trained to simulate how humans observe images, thus being more efficient and cheaper for the analysis of images. It may seem scary to think of the substitution of humans in the field of medicine, but the truth is that machines are being taught in every technological field. Therefore, Sect. 9.4 of this chapter also includes a basic explanation of how computers can be taught to carry out QA tests using model observers.

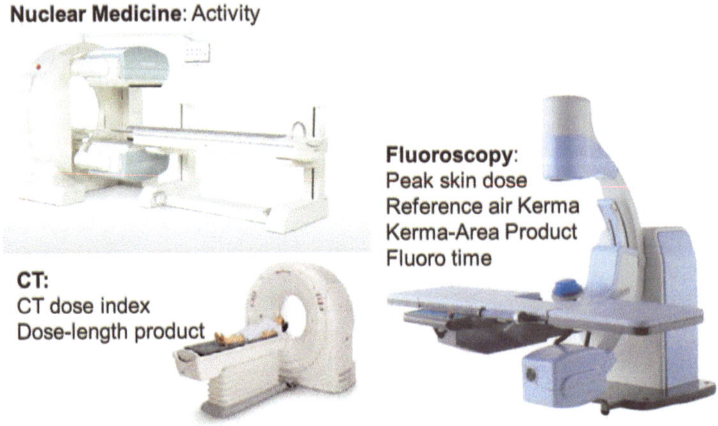

Fig. 9.3 Each imaging modality has currently its own dosimetry measures

This whole chapter focuses on current trends or new approaches to quality assurance. For standard procedures for dosimetry, radiation protection of the staff, and quality control of radiopharmaceuticals, the reader is referred to the previous volumes of the MADEIRA collection [9, 10] and to the very valuable compilation of information by the initiative Image Gently [11].

9.2 Dose-Related QA

Physicists working on any modality of medical imaging have developed different ways to evaluate the exposure of the patients to ionizing radiation, each fitting the corresponding technology. As a result, there is a large variety of such measures (Fig. 9.3). In nuclear medicine, the patient's dose depends on the activity that is administered as well as on the physical and biokinetic properties of the employed radiopharmaceutical. Consequently, it is the *activity* of the samples (in Bq) that is generally measured for quality assurance purposes. In radiography, the exposure is usually measured as *entrance surface air kerma* (ESAK) at the approximate position of the patient's skin. In the particular case of mammography, it is more common to measure the dose delivered at the breast's glandular tissue (*glandular dose*). For devices of fluoroscopy (such as the C-arms for interventional radiology and cardiology), four different measures are used; the maximum dose delivered at the patient's skin as peak ESAK or *peak skin dose* (PSD), or the air kerma measured at a particular spot along the central ray of the beam, called *reference air kerma*, or the product of the kerma at a certain position multiplied by the area irradiated, called *kerma-area product* (KAP), or the time that the device is recording a series of high-exposure images, called *fluoroscopy time*. Each of these measures has its advantages and drawbacks. In computed tomography (CT), ESAK and PSD have been used [12, 13], but the standard specific measures are the *CT dose index* (CTDI) and its product with the length of the scan, called *dose-length product* (DLP).

The problem with this large variety of different measures is that they are mostly noncomparable. Since they refer to certain definitions, each one must be measured under certain conditions and cannot be added up to other measures. For example, a certain CTDI measured after a CT scan cannot be added to the reference air kerma measured during a subsequent intervention to the same patient, even though they have the same units. Therefore, it is difficult to compare doses and prepare dose reports. The same problem appears during QA of recently marketed mixed modalities, such as PET-CT or SPECT-CT scanners, which combine nuclear medicine and conventional CT technology in a single machine; the radiation dose must be reported as two different measures.

As a solution to this situation, the current approach tends to homogenize dose measures using calculations of equivalent and effective dose. Equivalent and effective doses are defined for a reference individual and are calculated starting from absorbed doses in organs and tissues, taking into account different biological effectiveness of different types of radiation (by means of the radiation weighting factor w_R) and different sensitivities of organs and tissues to stochastic effects (by means of the tissue weighting factor w_T) [7]. In the context of medical exposures, these quantities are of value for comparing patient exposures from different diagnostic procedures or patient exposures using similar imaging procedures across different hospitals and different nations, and different imaging technologies for the same medical examination [14]. It should, however, be kept in mind that these quantities are reserved for use in risk assessment associated only with radiation-induced stochastic effects since the w_R and w_T values were calculated only in reference to these effects and not in relation to deterministic effects.

Because of the many calculations that are necessary to obtain equivalent and effective dose, it appears recommendable to additionally use a simpler quantity for QA purposes in cases of external exposures. The peak skin dose (PSD) is a perfect candidate because it is well understood, quite easy to measure, and directly relates to deterministic effects of radiation (such as the hair loss shown in Fig. 9.1). Furthermore, in transmission modalities (such as all X-ray examinations), it is at the patient's skin where the highest radiation doses are attained.

For this reason, current approaches to QA in terms of dose are focusing on measuring *organ doses* for the calculation of *effective dose* and on measuring *peak skin dose* for quick checks. Both quantities are not defined on the machines, but on the human body, which is the only common element of all modalities using ionizing radiation. Therefore, effective dose and skin dose can be easily compared and added up to account for examinations with different modalities.

9.2.1 Organ and Effective Dose

The measurements and calculations that are necessary to determine organ and effective doses have only recently started to be very accurate and reliable. Powerful computers, better tissue-equivalent phantoms, and smaller, more consistent detectors have made it possible. Computer simulations can be used to estimate how much dose a patient got after a certain exam, avoiding the need to carry out a whole dosimetry protocol for each patient.

Monte Carlo simulations for dosimetry purposes require the computation of many events per second because of the numerous collisions among photons, electrons, and nucleons. The collisions of the photons followed by the simulation determine the energy deposition in the surrounding tissue. Current studies suggest the use of a graphics processing unit (GPU) instead of a central processing unit (CPU). Using a GPU, the simulations can compute more than a million X-rays per second [15].

The simulation codes need to be validated experimentally. New tissue-equivalent phantoms have appeared that simulate the radiation attenuation characteristics of the human body almost perfectly (Fig. 9.4), and at the same time, more reliable detectors are marketed now that make measurements easier and more efficient.

As an example, the author and colleagues are currently using a flexible tissue-equivalent phantom, manufactured by the University of Florida (USA) and optically stimulated luminescence dosimeters (OSLDs). The phantom is made up of slices. Its gummy material allows the user to open inserts in the places where the dose is to be measured. Small detectors, such as OSLDs, can be inserted in those slots (Fig. 9.5).

The OSL readings obtained after a scan inform of the doses in the organ where they were placed, and those readings can be used to calculate the effective dose related to that particular examination [16] (Fig. 9.6).

Finally, another current development to introduce effective dose in the practice is a handbook of effective doses for all modalities using ionizing radiation. The project has recently started with X-ray computed tomography [17]. The object of this tool is to have the information handy when a patient or a member of the staff needs to know the amount of dose related to a certain examination.

In the future, these tools and methods may make it possible for patients to keep their own radiation record just as they keep their vaccination card.

9.2.2 Skin Dose

The thresholds for deterministic effects in the skin are well understood [18]. For example, erythema and temporary epilation appear after a minimum dose of 2 Gy [18]. Therefore, measurements of the maximum dose attained at the skin (or peak skin dose, PSD) are currently being suggested as a perfect candidate for regular checks of all X-ray devices for medical imaging, in particular CT scanners, which in general deliver the largest doses to the patients.

As an example, a group in collaboration with the author is measuring peak skin doses using radiochromic film. This kind of film turns dark when it is exposed to ionizing radiation, thus being appropriate for measurements of the beam width as well. Small pieces of film are enough for accurate, reliable measurements, but a calibration of the film batch is required. The darkening of the film after the irradiation can be measured using flatbed scanners or densitometers (Fig. 9.7, left).

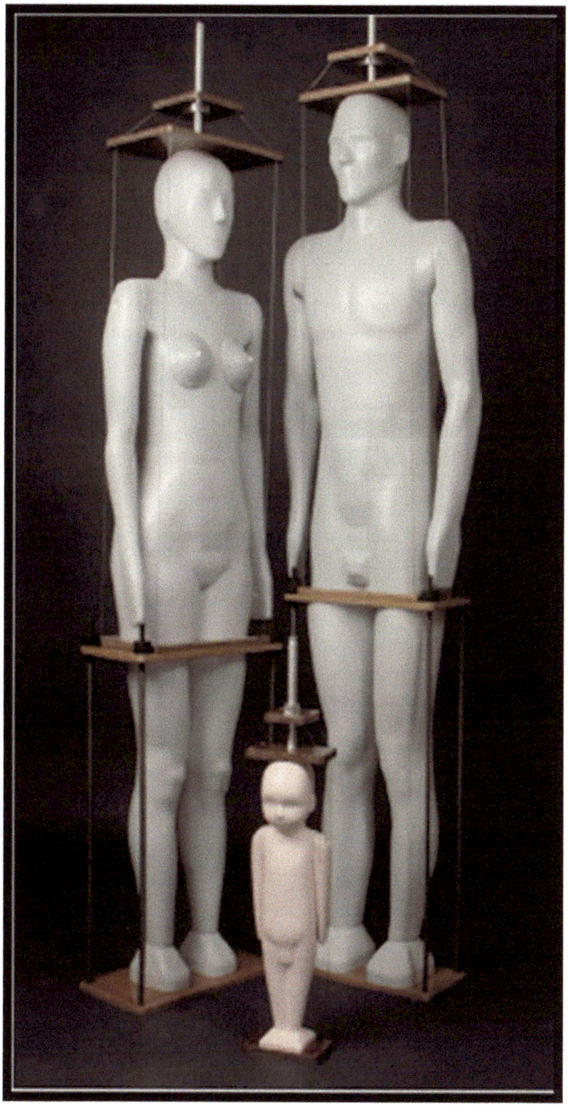

Fig. 9.4 Dosimetry phantoms. Photo from manufacturer's website

In our experiments, we have measured surface dose and CTDI at the same time to relate both quantities (Fig. 9.7, right). That way, technologists and doctors can instantly assess the PSD by looking at the CTDI displayed at the scanner's console.

Measurements like ours can serve in the future to develop and validate skin dose maps, which are especially important for high-dose examinations such as CT and interventional radiology. These maps display the received radiation field on the skin in a similar way to the level curves on an earth's surface map. The implementation of this feature in all devices would be a magnificent help for daily QA/QC purposes,

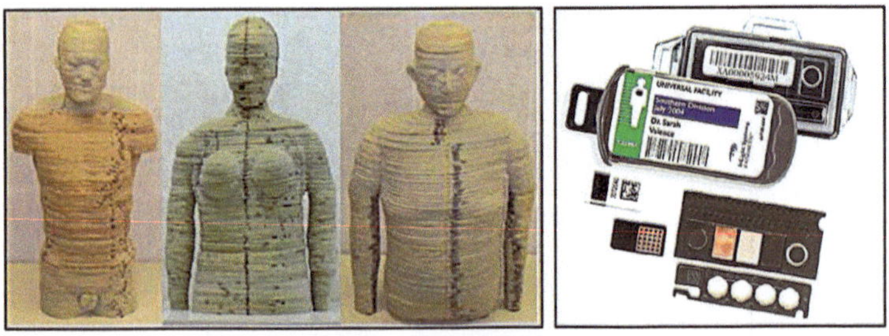

Fig. 9.5 The anthropomorphic phantom from the University of Florida (*left*), used by the author, and some OSL detectors (*right*). Photos from manufacturer's website

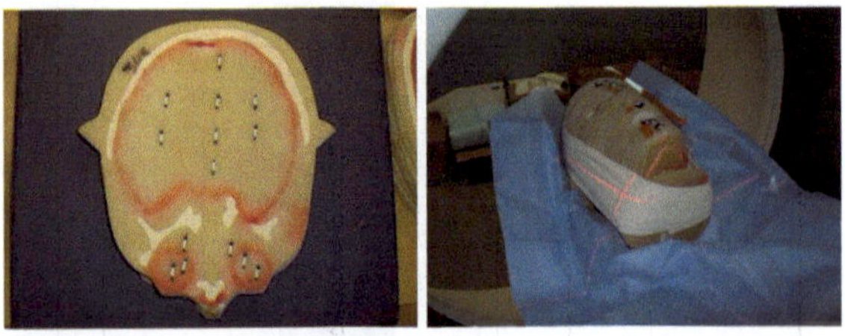

Fig. 9.6 A section of the UF phantom with inserted OSLDs (*left*) and the whole head inside a CT scanner (*right*)

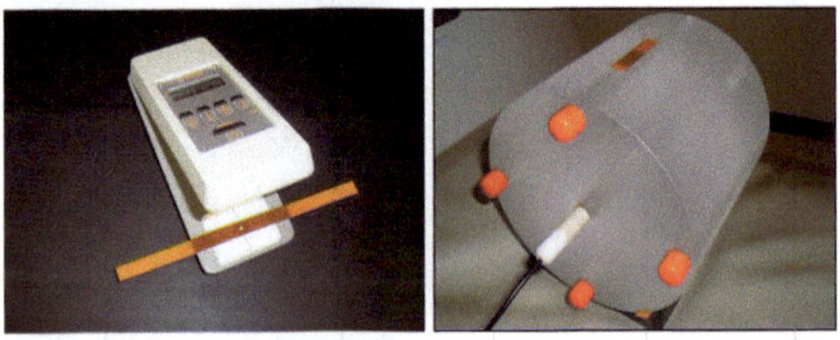

Fig. 9.7 (**a**) A densitometer for the determination of peak skin dose using film. (**b**) The conventional CTDI phantom with pieces of radiochromic film (*right*)

because doctors and patients would have access to the delivered dose and its location just after the procedure.

9.3 Image Quality Assurance

Image quality has been traditionally assessed by subjective methods; a few observers (doctors, other professionals, or volunteers) looked at a series of images and assessed which one looked better for a predefined task. For example, the same set of images containing a certain tumor and obtained with different image processing algorithms would be shown to several medical doctors. The doctors would be asked to rate the images depending on their ability to detect the tumor, and the best rated image would define which one of the algorithms is the best. These methods are still employed today, but they are very time-consuming and expensive, because a large number of observers and images are required to obtain reasonable conclusions. Furthermore, a similar study carried out in a different place may not be comparable just because the observers may be differently trained.

The alternative, standard approach is based on analyzing the resolution (or *high contrast resolution*) and noise (or *low-contrast resolution*) of sample images. Traditional methods use visual tests, whereas more recent methods involve measures based on Fourier analysis, such as the modulation transfer function (MTF) and the noise power spectrum (NPS). The drawbacks of these methods and some new alternative methods are the object of this section.

Another usual measure of image quality, the image contrast, is not so interesting from the point of view of QA, since this value can be modified by the user after the image has been taken by setting the center and the width of the window level. However, the contrast-to-noise ratio (CNR) or signal-difference-to-noise ratio (SDNR) is gaining importance in recent years and will be introduced in this section. Image artifacts (i.e., additionally displayed structures that confound the actual image of the object) are very interesting features from the point of view of QA, but there is such a wide variety of them for the different technologies that they had to be left out of the scope of this chapter. The reader is referred to specialized books for discussions on artifact detection and correction [19, 20]. At the end of this section, a brief discussion about the standard measure called signal-to-noise ratio (SNR) is given as a motivation for the last section of this chapter.

9.3.1 Noise

There are three main kinds of noise in medical images.

1. *Electronic noise* is the signal captured by a system in the absence of X-ray exposure. As an example, Fig. 9.8a shows the electronic noise pattern of a flat-panel detector. The stripped pattern is called synchronous or row-correlated noise. It may be due to the black currents inside the circuits or to the mechanism of image acquisition, which is made by synchronized readings of entire rows in the detector.
2. *Quantum noise* is the most traditional concept for the word "noise," that is, stochastic variations due to the quantum nature of the image pixels. Figure 9.8b shows an image with a large amount of this kind of noise. Note that any apparently homogeneous region actually contains a spectrum of different grey values within its pixels.
3. *Structural noise* is the image of everything that is present in the object but may confound the information that we are looking for, such as a tumor. Figure 9.8c shows the radiography of a chest including a lung tumor signaled by a triangular arrow. If the tumor had been a bit higher up, the structural noise due to the ribs would have made its detection much more difficult.

These kinds of noise together are what we refer to as "noise" in medical imaging. The contribution of electronic noise can be important in low-dose images, as required, for example, in digital breast tomosynthesis (DBT) [21], but it is usually

Fig. 9.8 Images showing an example of each kind of noise described in the text

insignificant. Structural noise can be decreased using appropriate settings in the imaging devices or using subtraction methods like in digital subtraction angiography (where background images are acquired separately and then subtracted from the actual radiographic images), but it is usually unavoidable.

Quantum noise is the most interesting source of noise, because it can be reduced by increasing the radiation dose to the patient, so that appropriate QA/QC is necessary to balance the two.

Quantum noise has been traditionally measured using the so-called pixel SNR, defined as μ_{sig}/σ, where μ_{sig} is the average grey value in a certain homogenous region of the signal and σ is the standard deviation of the grey value in a homogenous region of the background. The image in Fig. 9.9a has been obtained using a standard phantom in a CT device. In order to determine the SNR in this image, one would first have to decide which groups of four circles come out clearly above the background noise (and then calculate the corresponding SNR). This task is definitely very subjective, because some people would claim that they can *see* all five groups of four circles, but some others would even say that none of the five groups can be distinguished *clearly*! A similar situation happens in nuclear medicine using the Derenzo phantom (Fig. 9.9b), which was developed long ago [22] but it is still used and adapted for recent research studies [23, 24].

In addition, observe that the SNR is severely affected by the different offsets in the range of grey values chosen by different manufacturers.

A current trend suggested by several groups and supported by the author is calculating the contrast-to-noise-ratio (CNR) instead, defined as $(\mu_1 - \mu_2)/\sigma$, where the numerator measures the *difference* between the average grey values of a region in the signal and in the background [25]. Conti [26] have recently suggested using measures based on these quantities for the evaluation of image quality in time-of-flight (TOF) positron emission tomography (PET) scanners.

The use of the standard deviation as a measure of *homogeneity* is so widespread that it is found in highly innovative studies [27]. However, the use of the standard deviation to determine the *noise* of an image has a problem. Looking at the images of Fig. 9.10, we can see that their quality is very different or that they have different kinds of noise. However, these images have been obtained from the original on the left keeping the same value of the standard deviation as the original.

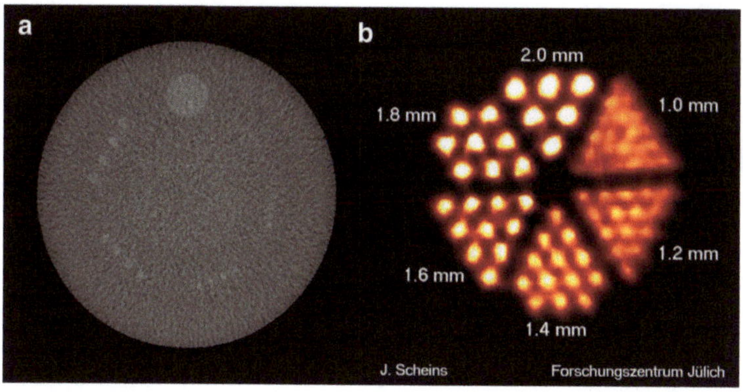

Fig. 9.9 (**a**) A CT slice of the GAMMEX 464 ACR accreditation CT phantom. (**b**) A slice of a mini Derenzo phantom obtained with a PET device (Courtesy of Raytest.de)

Therefore, using σ as a measure for noise, we would conclude that all four images have "the same noise"!

The reason for this mistake is easy to explain in the *Fourier* or *frequency domain*. If we take the Fourier transform of a homogenous image, we obtain its noise power spectrum (NPS). An image composed of tiny noise structures only (like the one in Fig. 9.10b) would show a large contribution of high frequencies to its NPS. On the contrary, an image composed of large structures only (like the one shown in Fig. 9.10d) would show a large contribution of lower frequencies in its NPS. These two cases are schematically represented in Fig. 9.11. You can see that the NPS is different in each case, but *its integral is the same*. This equal value of the integral is the reason for the standard deviation to be the same in all cases of Fig. 9.10.

Fourier-based methods like the NPS are widely extended now. However, they rely on some assumptions on the imaging system, such as shift invariance and linearity. The former means, for example, that the outskirts of the object are imaged exactly the same way as the center of the object. This is rigorously a wrong assumption, but it is acceptable. The latter assumption means that changes in the attenuation properties of the object would produce proportional changes in the corresponding grey values of the image, which is usually true but not always.

Therefore, there is a new trend to avoid Fourier-based measures and use image-based measures instead, such as the covariance matrix to study the noise. The covariance matrix K represents the correlation of one pixel to all others inside a certain region of interest (ROI) of an image. This correlation is assumed to be zero when using Fourier analysis. For a given number of ROIs of size $N \times N$, stretched as a vector of N^2 components $\mathbf{g}$, the covariance matrix K is calculated as

$$K_{ij} = \langle(\mathbf{g}_i - \bar{\mathbf{g}}_i)(\mathbf{g}_j - \bar{\mathbf{g}}_j)\rangle \quad i,j = 1,\ldots,N^2,$$

Fig. 9.10 Although each one of these images shows very different "noise" structures, they all produce the same standard deviation. Courtesy of Eduardo Guibelalde

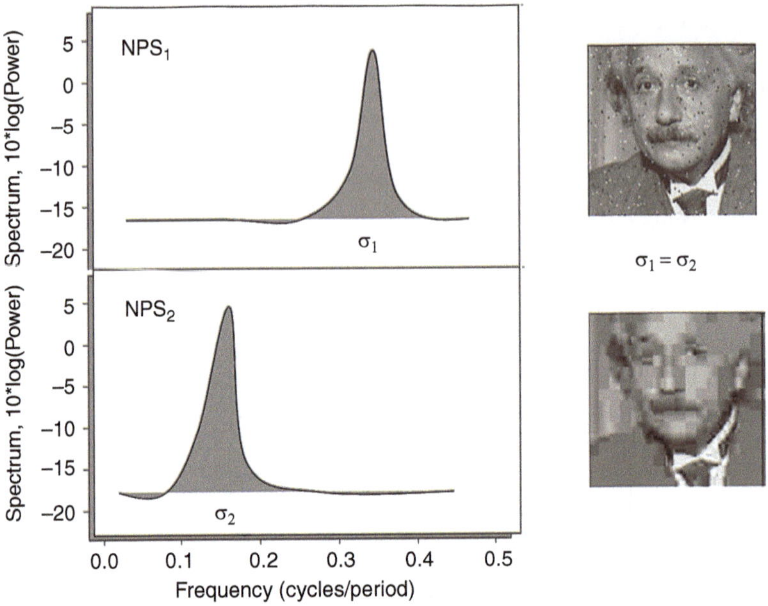

Fig. 9.11 The standard deviation is the integral of the noise power spectrum

where the angular brackets represent the average over all ROIs and $\overline{g}_i$ is the average value of the element i over all ROIs [28].

The matrix thus obtained is typically huge and not very easy to extract information from. Therefore, the covariance matrix can be conveniently averaged in a certain fashion and displayed as shown on the right of Fig. 9.12 [29]. This representation of the covariance matrix visually informs of the correlations in the image. The bright spot in the center represents the average correlation of each pixel to itself (which is always going to be the maximum of the covariance matrix), and the brighter zones represent pixel correlations in the same directions as those bright zones. In the case labeled patient, we can observe a more disperse correlation of pixels, whereas in the case labeled phantom, we observe a stronger correlation of pixels in the superior–inferior direction.

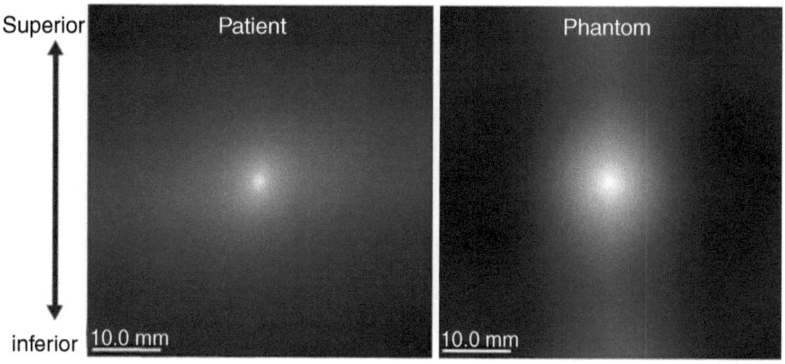

Fig. 9.12 A special representation of the covariance matrix, as used in [29], for two different cases (see text)

These images were taken from a recent evaluation study of a new phantom designed to be used both in X-ray as in MRI breast imaging modalities [29]. This phantom has been shown to possess very similar characteristics to breast tissue, and it could become an important tool for QA in the future.

9.3.2 Resolution

The traditional approach for QA of resolution is based on bar patterns as the one shown in Fig. 9.13. The observer (usually the physicist carrying out a regular check of a certain imaging device) must tell which group of five lines is clearly distinguishable. The problem with this kind of patterns is the same as the one described for noise; different people and even the same person in different moments would have different opinions about the same image. For example, I would say as I write this document that the resolution pattern corresponding to 22.4 lp/mm (line pairs per millimeter) is clearly resolved, but not the one corresponding to 25 lp/mm. Other observers may conclude that the group next to 31.5 is still resolvable, and both of us would have arguments to support our opinion.

A more objective method to measure spatial resolution is determining the full width at half maximum (FWHM) of the image of a linear object of a given size, as required by the NEMA standard document for gamma cameras [3]. As an example, this method has been used and adapted to measure the image quality for a recently developed SPECT collimator (Fig. 9.14) [30].

However, the quantitative Fourier-based method that is now becoming standard to measure spatial resolution is the modulation transfer function (MTF). This function informs about the ability of the imaging system to preserve the amplitude of a sinusoidal signal as a function of its frequency. The resolution of a system is better for a certain signal frequency when the MTF has higher values for that frequency. If we think of the signal as a row of pixels across each bar pattern in Fig. 9.13, this means that the MTF objectively contains, in a single graph, all the

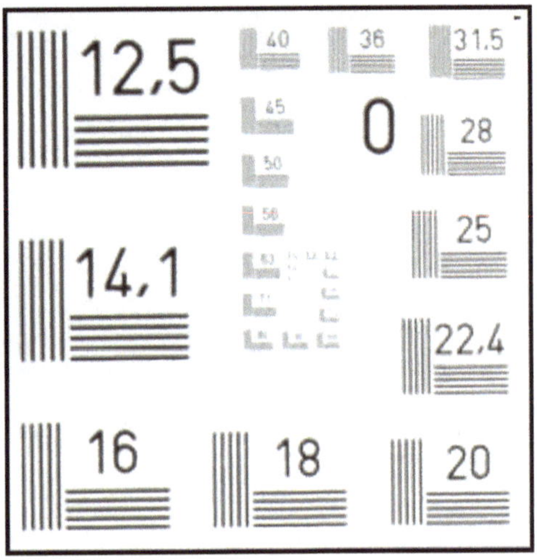

Fig. 9.13 A traditional bar pattern for the determination of high contrast resolution

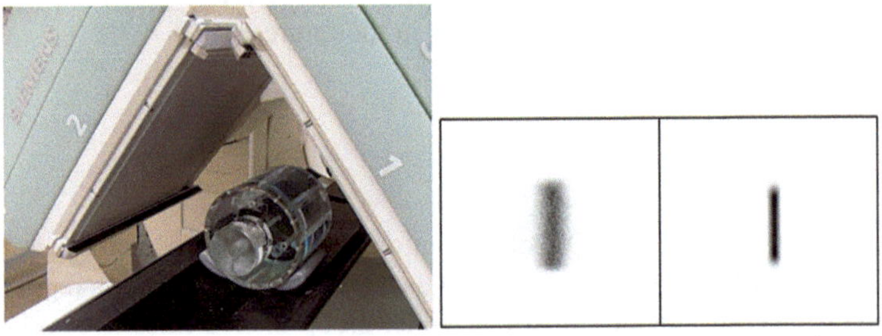

Fig. 9.14 (**a**) Experimental setup to measure the SPECT-reconstructed resolution in presence of scatter radiation [30]. A line source is placed inside the phantom made of PMMA. (**b**) Raw projection image of the line source. (**c**) Reconstructed sagittal slice

information that we can hardly obtain from many subjective analysis of diagrams like the one in Fig. 9.13.

In general, the amplitude of a low-frequency signal (related to the display of large objects) is easier to preserve for an imaging system, whereas preserving the amplitude of high-frequency signals (related to finer details in the object) is more difficult. Therefore, the MTF will always be close to one for low frequencies and will gradually decrease for higher frequencies. As an example, Fig. 9.15 shows results from a recent study using the MTF to analyze reconstruction algorithms in SPECT [31]. In particular, two iterative algorithms are compared to a standard reconstruction technique (black line).

A new approach to calculate MTF for CT scanners is currently being suggested to the International Electrotechnical Commission (IEC) [32]. The method is based on scanning a wire of a strong absorbing material (such as tungsten). The CT image of this wire's section is the so-called point spread function (PSF) of the system, and the

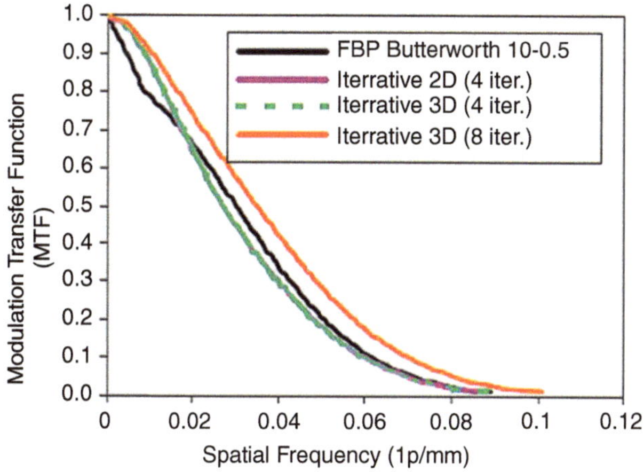

Fig. 9.15 A recent study [31] using the modulation transfer function for SPECT (see text). The *red line*, corresponding to a large number of iterations, shows the best spatial resolution at all frequencies

Fourier transform of the PSF is the wanted MTF of the system. Similar methods, based on calculating the MTF from a line spread function (LSF), have been recently suggested for single-photon emission computed tomography (SPECT), using a film-based flood source and materials easily found in a nuclear medicine department (Fig. 9.15) [31], as well as for positron emission tomography (PET) systems [33]. In this last study, the LSF is shown to be more appropriate than the PSF to calculate the MTF in nuclear medicine because the LSF involves greater data averaging.

9.3.3 The Signal-to-Noise Ratio

We have seen above that the traditional definition of pixel SNR hides all information about the frequencies of the noise. A different definition of SNR, named the Hotelling observer SNR [34], uses the ratio between the Fourier measures MTF and NPS seen above, thus preserving the frequency information. However, this definition of SNR has a different problem with current noise-reduction algorithms employed in some devices. These noise-reduction algorithms employ nonlinear image processing during the acquisition of images. This nonlinearity means that the filters used for homogenous regions (used to obtain the NPS) are different from the filters used for regions containing strong edges (such as the ones used to measure MTF). Therefore, at the moment of dividing MTF by the NPS, we are not taking into account the same image processing, and thus, the value obtained is not correct ("it's like dividing apples by oranges," in words of a colleague).

For this reason among others, another kind of QA tools is gaining a lot of attention in the past few years. These tools are called *mathematical or model observers*, because they simulate the ability of a human to detect information inside an image. Using model observers, the difficulties of the Fourier analysis and its assumptions are avoided, but of course, new difficulties arise from the need to accurately simulate human abilities. The last section of this chapter briefly explains how model observers may be used in the future for quality assurance purposes.

9.4 Opinion on the Future of Quality Assurance

The long-term goal for QA procedures in medical imaging is to gradually reduce and, if possible, avoid the need of humans for both QA purposes and diagnosis. This should not be surprising since every technological field shows the same trend. Robots and even androids are already taking over many tasks that were once carried out only by humans, because the use of robots is much cheaper and time-efficient, despite the associated sociological problems (such as the increasing unemployment rate and lack of personal contact). As an example, a robot[1] has recently been developed to carry out QA of X-ray devices through the calculation of the detective quantum efficiency (DQE, which is itself based on the Fourier measures MTF and NPS). The use of robots and computer-based tools has the additional advantage of avoiding individual subjectivity (i.e., the different response of humans under the same stimulus). Computer-assisted diagnosis (CAD) tools are already in use to help radiologists find tumors and other lesions.

In this last section, we explain how mathematical observers can work for QA purposes. First, we introduce a different tool for the analysis of image quality, the receiver operating characteristic (ROC) curves, which avoids the use of other image properties such as noise and resolution, and afterwards, we explain how mathematical observers can be trained to carry out their own ROC analysis.

In the future, QA of medical devices may be based on mathematical observers obtaining ROC curves and alerting the staff if the results are not satisfactory.

9.4.1 ROC Analysis

To perform an ROC analysis, say to evaluate the performance of a brain scan protocol in a new PET-CT device, the observer is shown two sets of images in arbitrary order; one set contains diseased cases (e.g., brain images from patients), and the other set contains non-diseased cases (e.g., brain images of healthy people). The observer must then assign a value, say from 1 to 10, to the likelihood that the imaged breast is healthy. That value is called "test result." If the task of finding the tumor is difficult enough, the test results will be distributed over two overlapping

[1] A small description of the robot can be found in the website of the manufacturer: http://www.imaging.robarts.ca/dqe/products.

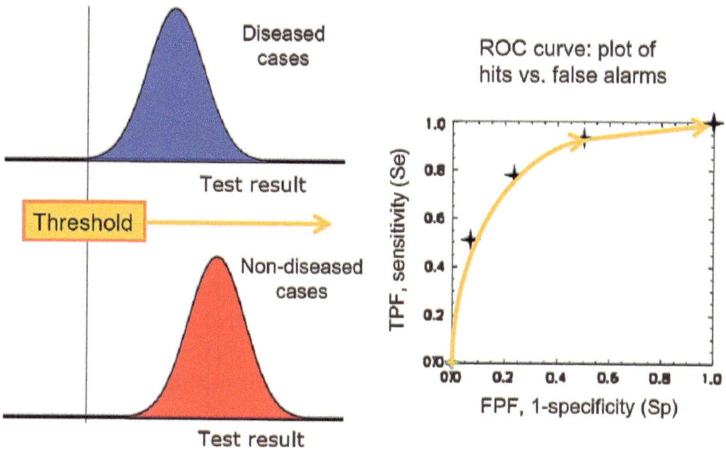

Fig. 9.16 Distributions of diseased and non-diseased cases. The ROC curve appears as the threshold is moved towards higher values of the test result. Courtesy of Kyle Myers

distributions; the distribution of the test results corresponding to diseased cases will be centered in a lower value than the distribution of test results corresponding to non-diseased cases (Fig. 9.16). The smaller the overlap of these two distributions, the better the ability of the observer to distinguish between the two cases will be, and the easier to reach a correct conclusion.

If we set a certain threshold, say 5, we may decide that all cases having a value below that threshold contain a tumor (positive cases), and the ones above that threshold do not (negative cases). In the usual case of overlapping distributions, some non-diseased cases (without tumor) will be labeled as diseased and treated as such (and will form the so-called false-positive fraction of cases, FPF), but also some diseased cases will be labeled correctly (and will form the true-positive fraction, TPF). The receiver operating characteristic (ROC) curve is just a plot of the FPF (false alarms) against the TPF (hits) when the threshold is varied from 0 to 10 (Fig. 9.16). As an academic example, the reader can think of what the TPF and FPF would be if the threshold is set to the extreme values 0 and 10.

ROC curves are used to assess the sensitivity and specificity of a certain device (or a certain observer under certain imaging conditions). The sensitivity refers to how many cases are actually detected, and the specificity refers to how well are diseased cases distinguished from non-diseased.

By the way, the solution to the academic example mentioned above is that if you set the threshold to a null value, all cases will be labeled as healthy (negative), so the TPF and the FPF will be zero (because no case was labeled as positive!). In the case of a high threshold value such as 10, both the TPF and the FPF take their maximum value. The optimum value for the threshold would be the one that produces the largest TPF at the smallest FPF possible.

So, how can one teach computers to carry out an ROC analysis?

9.4.2 ROC with Mathematical Observers

We can train computers to find tumors in medical images and to come up with a corresponding ROC analysis. The resulting tools of this training are called mathematical or model observers because they are designed to model human characteristics.

For a computer, an image is represented as a vector $\mathbf{g}$, and a lesion (such as a tumor) is for the computer a *signal*. If we obtain a large number of images containing a given signal $\mathbf{g}_1$, say brain or mammography images containing a small tumor, we can average them into an image $\overline{\mathbf{g}_1}$ that defines an average image of *background plus signal*. We can do the same with a large number of images $\mathbf{g}_2$ of *background only* (such as mammography images of healthy breasts) and average them into $\overline{\mathbf{g}_2}$. We can then represent a signal as

$$\text{Signal} \equiv \overline{\mathbf{g}_1} - \overline{\mathbf{g}_2}.$$

A computer can now obtain a test result for any image $\mathbf{g}$ by calculating the following scalar product:

$$\lambda = (\overline{\mathbf{g}_1} - \overline{\mathbf{g}_2})^t \mathbf{g}.$$

The term in brackets is the *training* of the observer, and the value λ obtained this way is the *observer*. If the input image $\mathbf{g}$ contains the signal that we had used in the images $\mathbf{g}_1$ (or something similar), the scalar product is going to be large, whereas if the input image does not contain the signal at all, the observer is going to produce a low value. By comparison of this value with a certain threshold, $\lambda_{\text{threshold}}$, we can obtain the necessary FPF and TPF for an ROC analysis.

Note that this is a very simplified case used for illustration purposes, because in order to train a computer to detect different kinds of signals, the training of the observer must contain other, more complex terms. This is an active field of research at the moment [35, 36].

In conclusion, QA in medical imaging is moving towards better simulation of human properties and abilities. Better anthropomorphic phantoms are being developed with improved detectors for dosimetry purposes, and IQ measures are being refined to account for the actual quality of the diagnostic information instead of looking at other purely physical properties. However, traditional methods have proved to be very practical and quite accurate, so that they will still be around for many years to come.

Acknowledgments For comments and support, I am deeply thankful to Kish Chakrabarti, Iacovos Kyprianou, Claudia Brunner (all working at Food and Drug Administration in Silver Spring, MD, USA), and Melanie Freed (currently working at the Pontificia Universidad Católica de Chile in Santiago, Chile). However, for possible incorrect facts stated in this chapter, I am entirely responsible.

References

1. Quality Assurance in Diagnostic Radiology. World Health Organization. Geneva (1982) As cited in "Radiation protection in Diagnostic and interventional Radiology" IAEA – Training Material on Radiation Protection in Diagnostic and Interventional Radiology. https://www.rpop.iaea.org/RPoP/RPoP/Content/index.htm
2. Performance Measurements of Positron Emission Tomographs (PETs) (2007) NEMA Standards Publication NU 2–2007. National Electrical Manufacturers Association, Rosslyn, VA
3. Performance Measurements of Gamma Cameras (3 December 2007) NEMA Standards Publication No. NU1. National Electrical Manufacturers Association (NEMA), Rosslyn, VA
4. Radiation overdoses point up dangers of CT scans. *New York Times*. 15 Oct 2009
5. Food and Drug Administration. Safety investigation of CT brain perfusion scans: update 11/9/2010. Accessed 20 Jan 2012. http://www.fda.gov/MedicalDevices/Safety/AlertsandNotices/ucm185898.htm
6. Gofman JW (1994) There is no safe threshold. Synapse (Journal of UCSF) 38(16):1–10
7. ICRP Publication 103 (2007) The 2007 Recommendations of the International Commission on Radiological Protection. ICRP publication 103. Ann ICRP 37(2–4):1–332
8. Brenner DJ et al (2003) Cancer risks attributable to low doses of ionizing radiation: assessing what we really know. Proc Natl Acad Sci USA 100(24):13761–13766
9. Cantone MC, Hoeschen C (eds) (2011) Radiation physics for nuclear medicine. Springer, Berlin
10. Mattsson S, Hoeschen C (eds) (2013) Radiation protection in nuclear medicine. Springer, Berlin
11. Image Gently site – radiation protection in nuclear medicine. http://www.pedrad.org/associations/5364/ig/?page=641
12. Avilés Lucas P, Dance DR, Castellano IA, Vañó E (2005) Estimation of the peak entrance surface air kerma for patients undergoing computed tomography-guided procedures. Radiat Prot Dosimetry 114(1–3):317–320
13. Joemai RMS, Zweers D, Obermann WR, Geleijns J (2009) Assessment of patient and occupational dose in established and new applications of MDCT fluoroscopy. Am J Roentgenol 192:881–886
14. Bolch WE, Eckerman KF, Sgouros G, Thomas SR (2009) MIRD pamphlet No. 21: a generalized schema for radiopharmaceutical dosimetry – standardization of nomenclature. J Nucl Med 50:477–484
15. Badal A, Badano A (2009) Accelerating Monte Carlo simulations of photon transport in a voxelized geometry using a massively parallel graphics processing unit. Med Phys 36(11):4878–4880
16. de las Heras H, Minniti R, Wilson S, Benevides L, Skopec M, Chakrabarti K (2011) Methods to determine peak skin dose and organ doses for computed tomography examinations poster at 53rd AAPM Annual Meeting, Vancouver, Canada. Abstract SU-E-I-46
17. Mills T, Stern S (2010) Development of a handbook of radiation doses in organs of patients undergoing X-ray computed tomography. Presentations at the 42nd national conference on radiation control, Newport, Rhode Island, 18–22 Apr 2010, and at the 52nd annual meeting of the american association of physicists in medicine, Philadelphia, PA, 18–22 July 2010
18. Balter S, Hopewell JW, Miller DL, Wagner LK, Zelefsky MJ (2010) Fluoroscopically guided interventional procedures: a review of radiation effects on patients skin and hair. Radiology 254(2):326–341
19. Hsieh J (2003) Computed tomography: principles, design, artifacts, and recent advances. In: Monographs, vol PM114. SPIE press, Bellingham, Washington, DC
20. Bushberg J et al (2002) The essential physics of medical imaging, 2nd edn. Lippincott Williams and Wilkins, Philadelphia, PA

21. Badano A, Kyprianou I, Jennings RJ, Sempau J (2007) Anisotropic imaging performance in breast tomosynthesis. Med Phys 34(11):4076–4091
22. Budinger TF, Derenzo SE, Gullberg GT, Greenberg WL, Huesman RH (1977) Emission computer assisted tomography with single-photon and positron annihilation photon emitters. J Comput Assist Tomogr 1(1):131–145
23. Park MA, Zimmerman RE, Tabener A, Kaye MW, Moore SC (2008) Design and fabrication of phantoms using stereolithography for small-animal imaging systems. Mol Imaging Biol 10(5):231–236
24. Zhou L, Vunckx K, Nuyts J (2011) Selection of point source configuration for multi-pinhole SPECT calibration. In: Proceedings of IEEE medical imaging conference. MIC18.M-16
25. de las Heras H, Schöfer F, Tiller B, Chevalier del Río M, Zwettler G, Semturs F (2012) An alternative method for QC in mammography and tomosynthesis. Proceedings of IRPA 13. Paper TS7d.5 and PO7.173
26. Conti M, Eriksson L (2011) Estimating image quality for future generations of TOF PET scanners. In: Proceedings of IEEE medical imaging conference. MIC 9.S-25
27. Rajaram R, Bhattacharya M, Ding X, Malmin R, Rempel TD, Vija AH, Zeintl J (2011) Tomographic performance characteristics of the IQ.SPECT system. In: Proceedings of IEEE medical imaging conference. MIC 9.S-64
28. Barrett HH, Myers KJ (2004) Foundations of image science, Ch. 14. Wiley, Hoboken
29. Freed M, Badal A, Jennings RJ, de las Heras H, Myers KJ, Badano A (2011) X-ray properties of an anthropomorphic breast phantom for MRI and x-ray imaging. Phys Med Biol 56:3513–33
30. Zeintl J, Rempel TD, Bhattacharya M, Malmin RE, Vija AH (2011) Performance characteristics of the SMARTZOOM collimator. In: Proceedings of IEEE medical imaging conference. MIC9.S-40
31. Fountos G et al (2009) Assessment of image quality in SPECT systems via the implementation of a novel flood source technique. In: Dössel O, Schlegel WC (eds) World Congress 2009, IFMBE Proceedings 25/II, pp 802–805
32. Brunner C, Renger B, Hoeschen C, Kyprianou IS (2011) Investigation of a method to estimate the MIF and NPS of CT towards creating an international standard. In: Proceedings of SPIE 7961. Medical Imaging 2011: Physics of Medical Imaging, 79613C (March 16, 2011)
33. Samartzis A et al (2011) Comparison of the image quality metrics performance in PET and SPECT imaging systems. In: Proceedings of IEEE medical imaging conference. MIC9.S-52
34. Liu H et al (2009) SKE/BKE task-based methodology for calculating Hotelling observer SNR in mammography, 72581D, SPIE medical imaging: physics of medical imaging
35. Hernandez-Giron I, Geleijns J, Veldkamp WJH, Calzado A (2011) Automated assessment of low contrast sensitivity for CT systems using a model observer. Med Phys 38(Suppl 1):S25
36. Zeng R, Petrick N, Gavrielides MA, Myers KJ (2011) Approximations of noise covariance in multi-slice helical CT scans: impact on lung nodule size estimation. Phys Med Biol 56 (19):6223–6242

Part III
New Concepts in Nuclear Medical Imaging

Chapter 10
The EURATOM FP7 Collaborative Project MADEIRA

Augusto Giussani and Marcus Söderberg

Part III of this book will give a closer insight into the outcomes of the European research project MADEIRA (Minimizing Activities and Doses by Enhancing Imaging quality in Radiopharmaceutical Administration) [1]. It was in the course of this project, co-funded by the European Commission through the EURATOM Seventh Research Framework Program from January 2008 till December 2010, that a training course on "Imaging in Nuclear Medicine" was organized (Munich, Germany, October 5–8, 2010) and this book was conceived. The course was the last in a series of three training courses organized in the frame of the project (http://www.madeira-training.org).

10.1 Context and Objectives of the Project

Six institutions of five European countries and one university from the United States of America were working in the project:

- Helmholtz Zentrum München (Neuherberg, Germany), which also coordinated and managed the project
- Medical Radiation Physics, Lund University, Skåne University Hospital Malmö (Malmö, Sweden)

A. Giussani (✉)
Research Unit Medical Radiation Physics and Diagnostics, Helmholtz Zentrum München – German Research Center for Environmental Health, Ingolstädter Landstr. 1, 85764, Neuherberg, Germany

Department of Radiation Protection and Health, BfS – Federal Office for Radiation Protection, Ingolstädter Landstr. 1, 85764, Oberschleißheim, Germany
e-mail: agiussani@bfs.de

M. Söderberg
Department of Clinical Sciences, Medical Radiation Physics Malmö, Lund University, Skåne University Hospital, SE-205 02 Malmö, Sweden

A. Giussani and C. Hoeschen (eds.), *Imaging in Nuclear Medicine*, DOI 10.1007/978-3-642-31415-5_10,

- Jožef Stefan Institute (Ljubljana, Slovenia)
- Instituto de Física Corpuscular (Valencia, Spain)
- Scivis wissenschaftliche Bildverarbeitung GmbH (Göttingen, Germany)
- Università degli Studi di Milano (Milano, Italy)
- University of Michigan (Ann Arbor, USA)

The starting point for the formation of the consortium was the consideration that three-dimensional (3D) functional nuclear medicine imaging, especially the combination of today's PET and SPECT systems with CT scanners, is a very powerful diagnostic tool that provides information about

- Three-dimensional regional distribution of the cancer and of its functional activity
- Three-dimensional functional activity of other (non-cancer) illnesses
- Type and staging of the disease to be treated in each specific case
- Outcome of the therapy

The use of these imaging procedures, which are based on the detection of the radiation emitted by (or traversing/scattered from) the patient, is inevitably associated with a sometime significant radiation exposure of the patient. The rapid growth of the individual dose due to medical exposures observed in the recent years calls for strategies aimed to minimize or at least reduce the patient's exposure. However, image quality, velocity, and easiness of the diagnostic examination are generally the main criteria leading to the introduction of new procedures.

Aim of the MADEIRA project, as its title reveals, was therefore to develop several strategies to improve the imaging process and thus enable to reduce the radiation dose to the patients while keeping the diagnostic information of the image at the same level or even enhancing it.

Three main approaches were implemented to achieve the goals of the project:

- Improvement of the imaging process by the development of a new add-on hardware consisting of a PET probe ring which can be inserted inside an ordinary PET or PET/CT system and enables to obtain detailed images of the region of interest
- Improvement of the imaging process by the development and implementation of new algorithms for noise reduction and image reconstruction, which are able to provide good quality diagnostic images even with reduced statistics
- Improvement of the diagnostic protocols and of the patients' dose evaluation for selected radiopharmaceuticals by extensive studies of their biodistribution and development of realistic biokinetic and dosimetric models

In addition to those scientific approaches, the MADEIRA project developed a large number of dissemination and education initiatives, among them the training courses mentioned at the beginning.

The main achievements of the project can be summarized as follows:

(i) The modeling, data gathering, and uncertainty determination methodologies developed within the framework of the MADEIRA project enabled to provide new insight into the biokinetics of selected medical nuclear imaging procedures, to calculate relevant dose distributions in the body and to optimize time schemes for better signal-to-noise ratios in the images [2–7].

(ii) The main concept and the feasibility of a new detection configuration, namely, the insertion of an add-on probe in a clinical PET system, could be successfully tested in a prototype. The predicted optimization of the resolution could be achieved and is ready to be implemented for clinically relevant procedures requiring improved spatial resolution [8–10].

(iii) New methods for testing the various image processing approaches applied to nuclear medical procedures were developed [11, 12]. Besides that, a new phantom for quality assurance and for effectiveness test in nuclear imaging was developed, and a patent application was filed. Another patent application was filed for a completely new reconstruction scheme for nuclear medical imaging based on compressive sensing techniques.

This chapter will describe mainly the achievements summarised in (iii). Some details on the compressive sensing techniques were already presented in Chap. 7. The following chapter will deal in detail with the add-on probe mentioned in (ii).

Putting the pieces together, the developments obtained in the frame of the MADEIRA project will provide possibilities to achieve images with higher spatial resolution and better signal-to noise ratios while decreasing at the same time the activity administered during nuclear medicine procedures, resulting in a reduction of the dose to the patient by a factor of three and even more [13].

10.2 New Algorithms for Image Reconstruction and Noise Reduction

In nuclear medicine imaging, the geometrical and statistical parameters are different compared to diagnostic radiology. On one side, the radiation source is inside the body, and one is interested to determine the original location of the emitted radiation as well as its intensity, whereas in X-ray-based tomography, the radiation source is outside the body, and one is interested to determine the anatomical structures by the way they absorb the incoming radiation. On the other side, the number of photons reaching the detectors in nuclear medicine is significantly lower than in projectional radiography, due to the presence of the collimators, necessary for identifying the origin of the emitted radiation. This means that the available reconstruction methods need to be adjusted to the specific case of nuclear medicine and that standardization is required for the evaluation process.

New algorithms and approaches were therefore developed within the MADEIRA project and then tested using images obtained from either clinical images of patients or, when not available, from experimental measurements with phantoms. Part of the work performed was facilitated by the software platform "scientific visualizer" developed by Scivis wissenschaftliche Bildverarbeitung

GmbH (Göttingen, Germany). This platform enables to read different image formats (Dicom and proprietary nuclear medicine image formats), to implement different image processing algorithms (e.g., OPED, ReSPECT) and to visualize and post-process the results. The possibility to decouple the different tasks by integrating different processing blocks as plug-ins facilitates the standardization of the evaluation of the results.

A completely new image reconstruction method for nuclear medical imaging was envisaged by scientists of Scivis GmbH and of the Helmholtz Zentrum München. This is based on the compressed sensing (CS) paradigm which allows for reconstructing data from very few measurements, provided the object is (in a mathematical sense) "sparse" or "compressible." The mathematical theory behind this strategy, as mentioned in Chap. 7, was however not able to deal with Poissonian noise which is characteristic of nuclear medicine imaging. Adapting the complex CS theory to Poissonian noise enables to reduce considerably the number of measurements necessary to produce the image while retaining the same image quality. A new methodology fulfilling these requirements was developed in the frame of the project and was filed as a patent application [14].

In addition, a structure-saving nonlinear noise reduction algorithm developed by Hoeschen et al. [15] was adapted to the nuclear medicine case (SPECT). This adaption was not straightforward, considering that the algorithm had been originally devised for reducing anatomical noise and statistical (i.e. mostly quantum) noise in projectional radiography and implemented into the scheme of CT investigations. It was necessary therefore to find optimal adaptive relocations to correct in particular for the high statistical noise in different imaging geometries typical of nuclear medical imaging.

The adaption to SPECT imaging was possible considering a SPECT instrument with the two detectors fixed at 180° to each other, each of them covering the whole 360° scanning field by rotation. Therefore, two data sets for correlation are available. However, the fact that the position of the emitting radioisotope is not always centered with respect to the rotation axis of the detectors requires a solution to take this shift into account. The depositioning implies indeed some time delay in the signal collection of the two detectors, and consequently a rotation phase shift has to be introduced for the data to be correlated.

A similar but even easier approach to reduce artifacts and the related noise in PET images was made possible by reconstructing two complementary data sets which are then added together. With this procedure, undersampling artifacts are avoided, as shown in Fig. 10.1. The images shown in panels (a) and (b) are two different reconstructions of the same phantom done according to standard sampling schemes. The combination of the reconstructions using complementary data sets enables to eliminate the artifacts, as observed in panel (c) of the image.

The algorithms developed in the course of the project were tested by assessment of reconstructed images using the "scientific visualizer" tool.

A number of measurements with a Jaszczak phantom (Data Spectrum Corporation, Model # 5000) were performed to study how different reconstruction parameters influence the image quality, in particular to find the optimal number

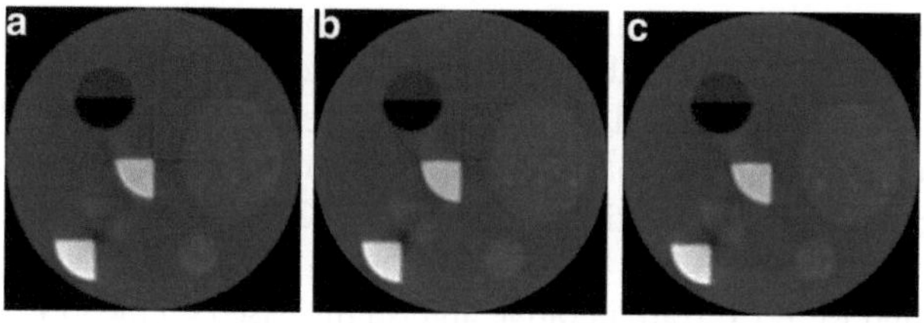

Fig. 10.1 Reconstruction of a phantom slice using two different sampling schemes which are supplementary to each other (panels (**a**) and (**b**), respectively). In the standard reconstruction, this would still produce the same artifacts as seen here. The summation of the reconstructions of the complementary data sets is shown in part (**c**) of the figure. The aliasing artifacts are destroyed due to the principal accuracy of the OPED algorithm

of subsets and iterations required to be used. In addition to the phantom images, also raw data for patients suffering from neuroblastoma or pheochromocytoma and investigated with ^{123}I-metaiodobenzylguanidine (MIBG) were collected. Images were reconstructed for different equivalent number of iterations (defined as the number of subsets times the number of iterations), using the Siemens iterative algorithm Flash 3D. The algorithms ReSPECT (iterative) and OPED (analytic) were adapted for ^{123}I-MIBG examinations and implemented in the "scientific visualizer" platform. Details of this analysis are given in the following section.

Also the new noise reduction methods developed in the MADEIRA project and used in combination with OPED were implemented in ReSPECT. Denoised patient images were reconstructed with ReSPECT and OPED.

10.3 Rank-Order Study

No current image quality criteria are established for nuclear medicine examinations as there are for X-ray images. ROC studies or derivatives thereof are widely used in medical X-ray imaging to define differences in imaging procedures, but ROC studies are time-consuming and require a large patient cohort with normal and pathological subjects, and the right diagnosis for each case should be known. However, if the diagnostic performances are comparable or only slightly different, performing a side-by-side review (rank-order study) may be desirable [16, 17]. In the frame of the MADEIRA project, such a rank-order study was performed for SPECT imaging with ^{123}I-MIBG [18].

SPECT is indeed afflicted with relatively poor spatial resolution and high statistical noise compared to other medical imaging systems. This limitation is partly explained by the restricted amount of reasonable activity, photon attenuation within the patient, scatter, reasonable acquisition time, light scatter in the detector, detector photomultiplier tube configuration, and the trade-off between efficiency and the spatial resolution of the collimator. By choosing an appropriate reconstruction method and optimal reconstruction parameters, opportunities exist for improving image quality and lesion detectability.

To find the optimal conditions for a reconstruction algorithm, several parameters need to be optimized. The number of equivalent iterations (EI) was optimized for the Flash 3D (Siemens Medical Solutions, Forchheim, Germany) [19] reconstruction algorithm and compared to two recently developed reconstruction algorithms, ReSPECT (Scivis GmbH, Gottingen, Germany) [20] and orthogonal polynomial expansion on disc (OPED) [12, 21].

Flash 3D and ReSPECT are iterative algorithms based on the OSEM technique, and OPED is an analytic algorithm. A rank-order study was performed, and the SPECT images were interpreted by three experienced observers, all nuclear medicine physicians, by showing the image sets in the software package "scientific visualizer" installed on one of the department's regular PACS workstations. The software was adapted for observer studies with the possibility of showing up to eight unlabeled image sets side by side.

The images used for the study were obtained for 11 patients who had underwent SPECT 4 h after intravenous injection of approximately 200 MBq. Additionally, images acquired for 14 patients at 24 h postinjection (p.i.) were also included in the study. The SPECT data for each patient were presented side by side in sagittal, coronal, and transversal views and as a maximum intensity projection in random order and unlabeled. The images were presented in a black-and-white scale, and the observers were free to change the window-level settings. No time limit was imposed on the observer's evaluation. In the first visual assessment, images reconstructed at eight different EI numbers for Flash 3D were displayed (8, 16, 32, 64, 80, 96, 128 and 256 EI). It should be noted that using too few EI is undesirable as the algorithm may not reach convergence everywhere in the reconstructed volume and lesions will not be visible.

Three observers were asked to rank the three best image sets according to their overall impression of the image quality with regard to noise level, ability to discriminate uptakes in anatomical structures (e.g., liver, adrenal glands, kidneys, and spleen), introduction of artifacts, and if possible delineation of suspected pathology. The rank order was 1 (best) to 3, and the remaining image sets obtained a rank order of 4. This procedure was performed for images acquired 4 h and 24 h after injection. The average distribution of image quality ranking for all observers was calculated for the different EI numbers. In addition, the rank order was obtained based on the calculated average of scores a given image set received, i.e. the lower the value the better the image. In the second visual assessment, the two best considered EI numbers for Flash 3D were compared to ReSPECT and OPED (Fig. 10.2).

In the same manner as the first assessment, the observers were asked to rank the image sets according to their overall impression of the image quality. A rank order of 1 was assigned to the image set judged to have the best overall image quality, and a rank of 4 was assigned to the worst one. The procedure was performed for images acquired 4 h and 24 h after injection. The average distribution of image quality ranking for all observers was calculated for the different reconstruction methods. In addition, the rank order was obtained based on the calculated average of scores that a given image set received.

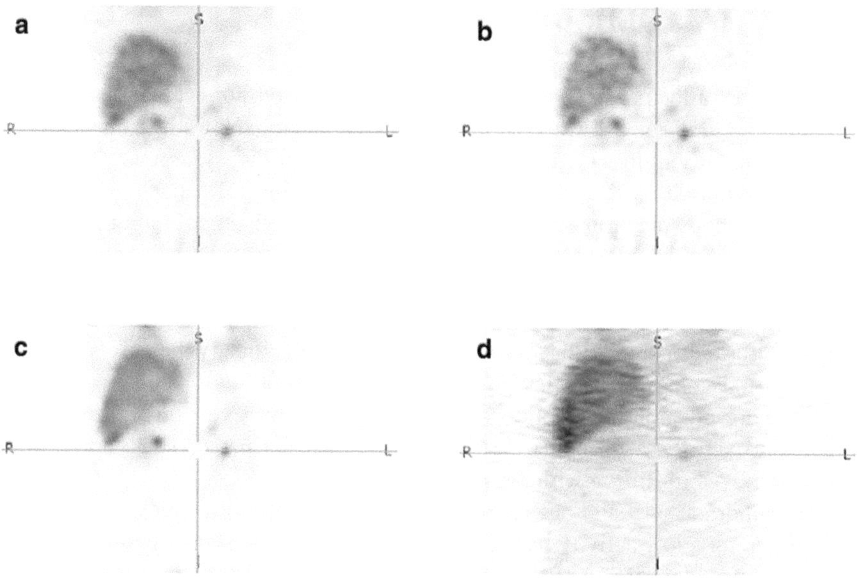

Fig. 10.2 Coronal images acquired 4 h after the injection of approximately 200 MBq ^{123}I-MIBG and reconstructed using different reconstruction methods: Flash $3D_{32}$ (**a**), Flash $3D_{64}$ (**b**), ReSPECT (**c**), and OPED (**d**)

The optimal EI number for Flash 3D was found to be 32 for both acquisitions at 4 h and 24 h p.i. The average rank order for the different reconstruction algorithms was Flash $3D_{32}$ > ReSPECT > Flash $3D_{64}$ > OPED for acquisitions at 4 h. Similarly, for acquisitions at 24 h, the order was Flash $3D_{16}$ > ReSPECT > Flash $3D_{32}$ > OPED.

The algorithm developed the frame of the project shows apparently still potential for improvement. It should however be considered that neither attenuation nor scatter correction was applied in the OPED algorithm. The reconstructed images using OPED were noisy and had streaklike artefacts due to the geometry of the SPECT data. Actually, OPED is more suitable for use in PET and CT [22], since it requires sinusoidal lateral sampling. Instead, the measured SPECT data are parallel in the classical sense, i.e. uniformly distributed projections with equi-spaced lateral sampling, due to the collimation used, and consequently, they need to be resampled for an optimal elaboration with OPED.

The results obtained in this study, although specific for the scanner and parameter settings and depending on the amount of radionuclide activity used, acquisition parameters, acquisition time, and examined body area, give however an indication of preferred reconstruction parameters and algorithms, which need to be further investigated with a larger patient cohort. Therefore, they are useful for future optimization of reconstruction methods and parameter settings.

10.4 The MADEIRA Phantom

In the course of the project, the need of a specific phantom for evaluation of spatial resolution in PET and SPECT imaging was acknowledged. Nowadays, the majority of phantoms used to characterize a SPECT or a PET system are cylinders containing fillable inserts, e.g., the Jaszczak PET/SPECT phantom [23] and NEMA IEC body phantom setTM [24], or phantoms consisting of multiple discrete disks or sheets, e.g., the Hoffman phantom [25] and the porous phantom [26]. Tissue-equivalent anthropomorphic phantoms, as the RSD Alderson heart/thorax phantom, are also available [27]. These phantoms however cannot give detailed and precise information regarding spatial resolution and specifically the onset of the so-called partial volume effect (PVE).

An ad hoc phantom was therefore devised. A preliminary version of the phantom (named MADEIRA phantom) (see Fig. 10.3, left) consisted of 21 cones with separate valves for filling with radioactive solution. The cones were placed within a half-cylindrical surrounding vessel, which is also fillable with radioactive solution for simulating a low activity background. The phantom was constructed of acrylic glass for visual inspection of bubble-free filling.

The phantom was developed by Scivis in cooperation with the Medical Radiation Physics Department in the Skåne University Hospital Malmö as a clinical partner. It was designed to specifically fit into the RSD Alderson thorax phantom, so that it could be used for a standardized testing in a clinically relevant situation.

Preliminary tests highlighted some construction and leakage problems, which were corrected in the second version of the phantom, consisting of 16 cones of 19 cm length and with an inner diameter linearly decreasing from 16 mm to 2 mm. The wall thickness of the cones is 1 mm. Ideally, the cone walls should have non-zero thickness to prevent the so-called wall effect, i.e. the fact that a shell of zero activity divides the cone solution and the background solution. This may affect quantitative measurements and evaluation of lesion detection, but it is believed to be negligible concerning the evaluation of PVE. The chosen thickness of 1 mm was a practical compromise to guarantee the fabrication of the phantom and at same time to minimize the "wall effect".

The phantom was employed for measurements in a clinical environment. SPECT (^{99m}Tc) and PET (^{18}F) measurements were performed, with the cones filled with activity levels differing by a factor of 3/4 from one to the other. The external vessel was filled with a background activity concentration which was one-tenth of the concentration in the most active cone and ten times the concentration in the less active cone. Consequently, eight cones contained activity below background and eight cones activity above background. Initial activities at the start of acquisition were 530 MBq for ^{99m}Tc and 145 MBq for ^{18}F.

The SPECT acquisitions were performed with a Symbia T2 SPECT/CT (Siemens Medical Solutions, Forchheim, Germany) in noncircular orbit step and shoot mode. The collimator used was a low-energy ultrahigh resolution (LEUHR) [28]. The applied energy windows were a 15 % photo-peak window symmetrically

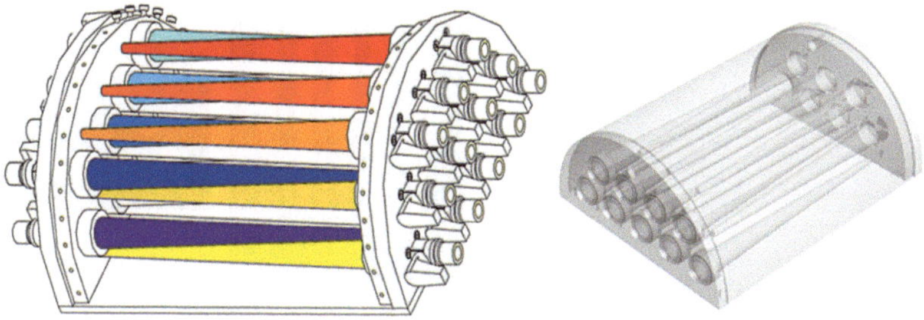

Fig. 10.3 The MADEIRA phantom. *Left*: first design; *right*: design used for the tests shown here

placed around 140 keV and a 15 % lower scatter window. Each projection was acquired for 60 s, and 240 projections were acquired over 360°. The acquisition matrix was 256 × 256. Images were reconstructed using the iterative algorithm ReSPECT 3.0, using eight iterations, nonlinear noise suppression and without Gaussian filter (Scivis wissenschaftliche Bildverarbeitung GmbH, Göttingen, Germany).

The number of projections was varied by taking only a subset out of the total projections, preserving an equidistant sampling of the angle space (120, 80, 60, 48, 40, 30, and 24). The total number of photons was varied by Monte Carlo simulations (increasing Poisson noise), intended to virtually simulate shorter acquisition time or activity. The matrix size was varied by downsampling the projection matrix (256, 128, 64). This made it possible to investigate the influence of count statistics, number of angles and matrix size simultaneously by just one (real) acquisition. All manipulations were carried out by using the Scivis "scientific visualizer" software, producing correct DICOM data sets that can be reconstructed by any DICOM capable software. The PET acquisitions were performed with a Gemini 16 PET/CT (Philips Healthcare, Best, The Netherlands) using a matrix size of 144 × 144 and a scan time of 15 min per bed position. The iterative reconstruction method used was LOR-TF-RAMLA, with default settings. Further details on the experiments can be found in [29].

Figure 10.4 shows representative SPECT and PET images of the new phantom performed with very good statistics. Spatial resolution was evaluated as the full width at half maximum (FWHM) by drawing profiles across the center of the cones.

Figure 10.5 shows one application of the phantom for the evaluation of spatial resolution and, in particular, of the partial volume effect. As explained in Chap. 7, PVE is manifested in the underestimation of the activity contained in a region, when the dimension of this region falls below a certain threshold (according to Skretting [30], the phenomenon is better described by the term "intensity diffusion"). The contrast-to-noise ratio (CNR) of lesion to background is essential for the detection of lesions in SPECT and PET images. Below a certain volume, the reconstructed intensity tends to diffuse into neighbouring voxels, thereby resulting in very low target-to-background ratios (underestimation of activity concentration in the target) [31]. The MADEIRA phantom has been designed so that it can simultaneously provide different target-to-background activity ratios with a linearly

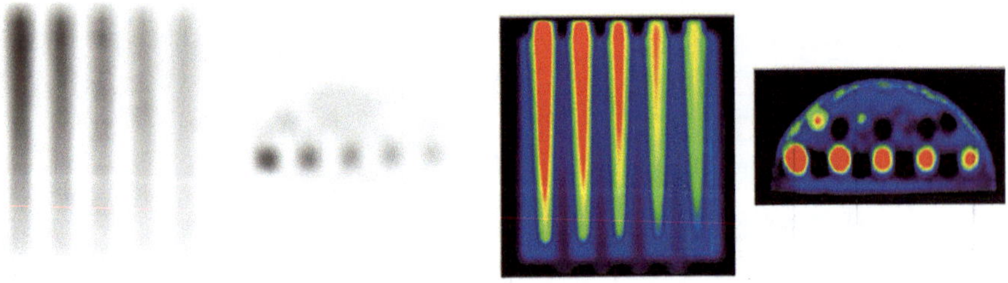

Fig. 10.4 Representative SPECT (*left*) and PET (*right*) images of the new phantom

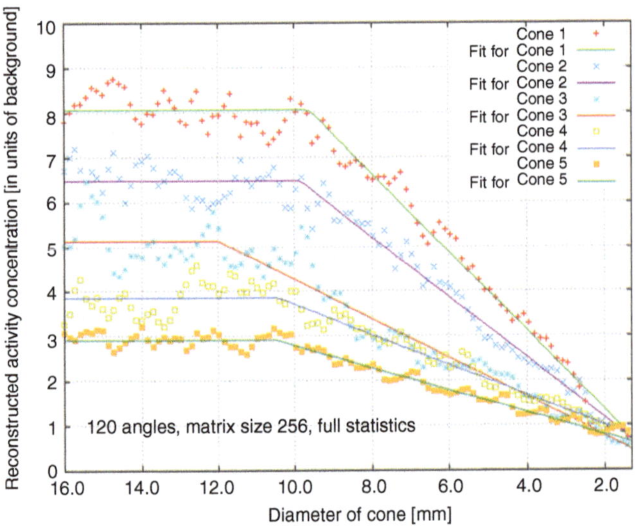

Fig. 10.5 Reconstructed activity concentration as a function of the cone diameter as resulting from the analysis of the SPECT images of Fig. 10.4. The activity concentration was normalized to the background activity concentration, measured in a cone-free area inside the vessel

changing diameter of active or inactive lesions. It was thus possible to evaluate PVE by drawing profiles along the length of the cones. The use of a phantom with cones, which change their diameters continuously and not stepwise, is actually instrumental in finding the dimension of the region when the PVE starts to play a role. Figure 10.5 shows that for SPECT images, this happens when the diameter is about 1 cm or less.

The level of reconstructed activity concentration in the constant part was not affected by the number of projections used for reconstruction. On the contrary, the number of projections used for reconstruction affected the value of the cone diameter where PVE starts: the lower the number of projections, the larger the size of the objects affected by PVE [29]. Consequently, care should be taken in the reconstructing images with a reduced number of projections, since this reduction will increase the importance of PVE.

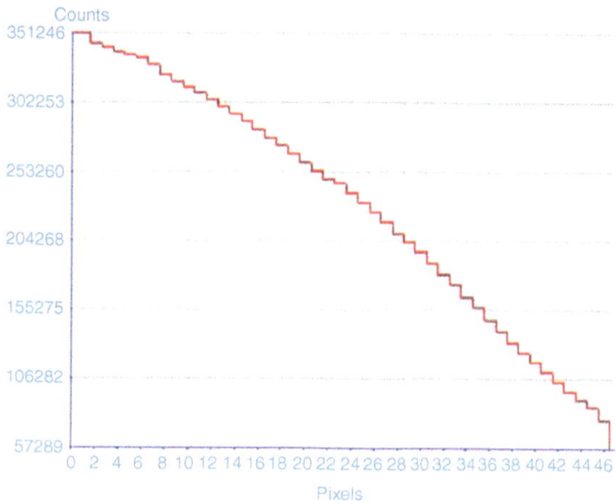

Fig. 10.6 Reconstructed activity concentration as a function of the cone diameter as resulting from the analysis of the PET images of Fig. 10.4

A different pattern was observed in the analysis of the PET images (Fig. 10.6). No plateau was observed before the starting point of the PVE. This finding was somewhat surprising, a better spatial resolution for the PET system compared to the SPECT system was actually expected. The results indicated that the onset of the PVE begins already at sizes larger than the bases of the MADEIRA phantom cones. It should be taken into consideration that the noise suppression and regularization mechanisms chosen by the authors of the study and included in the reconstruction algorithms could have affected the results.

Other applications of the phantom included the evaluation of figures of merit such as detectability, signal-to-noise, and resolution depending on the total number of counts and projections and parameters for various reconstructions. In order to do that, further tests were performed by varying the acquisition parameters and the count statistics for the SPECT measurements. By using Monte Carlo simulations, the total number of photons was varied. The number of projection angles was varied by taking only a subset out of the 240 total projections. The matrix size was reduced by rebinning.

Figure 10.7 shows the result obtained in the reconstruction of the MADEIRA phantom. The left panel gives the reconstructed images using the compressed sensing (CS) methodology with only 15 projection angles. Note the smoothness of the hot cones and the detectability of the cold ones. The image in the right panel of Fig. 10.7, obtained with standard algorithms from 48 projections (i.e., more than three times the ones used with CS), appears slightly less noisy; however, the hot cones have jagged edges, and the cold cones are hardly discernable at all. Thus, it can be clearly stated that CS turns out to be extremely advantageous to reduce to about one-third the amount of projections necessary for artifact-free reconstruction or alternatively to reduce the necessary activity administered.

The MADEIRA phantom has thus proved its potential as an important practical tool for comparison and optimization of different acquisition and reconstruction parameters in nuclear medicine tomographic studies. It can be employed to find the best working point of a given system as well as for comparisons between various tomographic units.

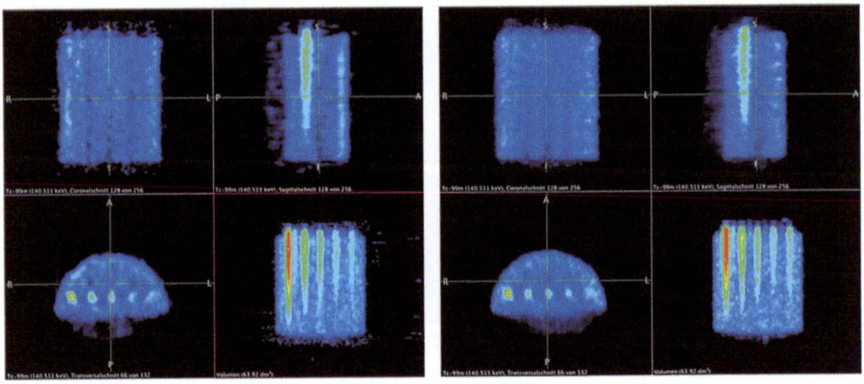

Fig. 10.7 *Left*: Reconstruction of the MADEIRA phantom from *15 angles with very poor statistics* (23 counts maximum) *using CS*. *Right*: *Conventional penalized likelihood reconstruction* from *48 angles* (more than three times the number of angles of the CS reconstruction)

A further improved version of the phantom, with a reduced number of cones (12) and two emptying valves for facilitating its cleaning, will be commercially available starting end of 2012.

10.5 Final Considerations

In this chapter, some of the most relevant outcomes obtained in the frame of the European collaborative Project MADEIRA with regard to image acquisition and processing procedures were presented. Together with the development of new detection systems, like the one described in the next chapter, the application of such procedures can lead to a significant reduction of the activity administered to the patients undergoing nuclear medical diagnostic procedure and consequently a significant positive impact on their exposure to ionizing radiation (and specifically the dose to the healthy tissues) while guaranteeing or even improving the quality of the diagnostic information provided.

Acknowledgments The authors would like to thank Gernot Ebel for useful comments and suggestions during the preparation of this contribution.

References

1. Hoeschen C, Mattsson S, Cantone MC, Mikuz M, Lacasta C, Ebel G, Clinthorne N, Giussani A (2010) Minimizing activity and dose with enhanced image quality by radiopharmaceutical administrations (MADEIRA). Radiat Prot Dosimetry 139:250–253
2. Sydoff M, Uusijärvi H, Leide-Svegborn S, Mattsson S (2010) Absolute quantification of activity content from PET images using the Philips Gemini TF PET/CT system. Radiat Prot Dosimetry 139:236–239

3. Janzen T, Tavola F, Giussani A, Cantone MC, Uusijärvi H, Mattsson S, Zankl M, Petoussi-Henß N, Hoeschen C (2010) Compartmental model of ^{18}F-choline. SPIE Meeting, San Diego, CA
4. Uusijärvi H, Nilsson L-E, Bjartell A, Mattsson S (2010) Biokinetics of ^{18}F-choline studied in four prostate cancer patients. Radiat Prot Dosimetry 139:240–244
5. Tavola F, Janzen T, Giussani A, Facchinetti D, Veronese I, Uusijärvi-Lizana H, Mattsson S, Hoeschen C, Cantone MC (2012) Nonlinear compartmental model of ^{18}F-choline. Nucl Med Biol 39:261–268
6. Giussani A, Janzen T, Uusijärvi-Lizana H, Tavola F, Zankl M, Sydoff M, Bjartell A, Leide-Svegborn S, Söderberg M, Mattsson S, Hoeschen C, Cantone MC (2012) A compartmental model fpr biokinetics and dosimetry of ^{18}F-choline in prostate cancer patients. J Nucl Med 53(6):985–993
7. Li W, Hoeschen C (2010) Uncertainty and sensitivity analysis of biokinetic models for radiopharmaceuticals used in nuclear medicine. Radiat Prot Dosimetry 139:228–231
8. Linhart V, Burdette D, Chessi E, Cindro V, Clinthorne NH, Cochran E, Grošičar B, Honscheid K, Kagan H, Lacasta C, Llosá G, Mikuž M, Stankova V, Weilhammer P, Žontar D (2011) Spectroscopy study of imaging devices based on silicon Pixel Array Detector coupled to VATAGP7 read-out chips. J Inst 6:C01092
9. Studen A, Burdette D, Chesi E, Cindro V, Clinthorne NH, Cochran E, Grošičar B, Honscheid K, Kagan H, Lacasta C, Llosá G, Linhart V, Mikuž M, Stankova V, Weilhammer P, Žontar D (2009) Performance of the MADEIRA PET probe prototype. 2009 NSS/MIC IEEE conference, Orlando, FL
10. Studen A, Burdette D, Chesi E, Cindro V, Clinthorne NH, Cochran E, Grošičar B, Kagan H, Lacasta C, Linhart V, Mikuž M, Stankova V, Weilhammer P, Žontar D (2010) Timing performance of the silicon PET insert probe. Radiat Prot Dosimetry 139:199–203
11. Brzezinski K, Gillan J, Oliver JF, Lacasta C, Rafecas M (2011) Multiple resolution PET data: image properties using the list-mode ML-EM reconstruction algorithm. 11th international meeting on fully 3D image reconstruction in radiology and nuclear medicine, Potsdam, Germany
12. Tischenko O, Xu Y, Hoeschen C (2010) Main features of the tomographic reconstruction algorithm OPED. Radiat Prot Dosimetry 139:204–207
13. MADEIRA Final report. Available at www.helmholtz-muenchen.de/en/madeira/madeira-home/index.html
14. Aspelmeier T, Ebel G, Hoeschen Ch. Tomographic imaging using Poissonian detector data. International Patent Application PCT/EP2010/000526
15. Tischenko O, Hoeschen C, Buhr E (2005) An artifact-free structure saving noise reduction using the correlation between two images for threshold determination in the wavelet domain. In: Fitzpatrick JM, Reinhardt JM (eds) Medical imaging 2005: image processing. Proc SPIE 5747:1066–1075
16. Towers JD, Holbert JM, Britton CA, Costello P, Sciulli R et al (2000) Multipoint rank-order study methodology: observer issues. Invest Radiol 35:125–130
17. Good WF, Sumkin JH, Dash N, Johns CM, Zuley ML et al (1999) Observer sensitivity to small differences: a multipoint rank-order experiment. Am J Roentgenol 173:275–278
18. Söderberg M, Mattsson S, Oddstig J, Uusijärvi-Lizana H, Valind S, Thorsson O, Garpered S, Prautzsch T, Tischenko O, Leide-Svegborn S (2012) Evaluation of image reconstruction methods for ^{123}I-MIBG-SPECT: a rank-order study. Revision requested by Acta Radiol
19. Hawman E, Vija AH, Daffach R, Ray M (2003) Flash 3D technology optimizing SPECT quality and accuracy. Whitepaper. Siemens Medical Solutions, Forchheim, Germany, pp 1–6
20. Scivis GmbH (2006) ReSPECT – technical description. Version 2.0. Scivis, Gottingen, Germany, pp 1–27
21. Xu Y, Tischenko O, Hoeschen C (2007) Image reconstruction by OPED algorithm with averaging. Numer Algorithms 45:179–193
22. Xu Y, Tischenko O, Hoeschen C (2006) A new reconstruction algorithm for radon data. Proc SPIE 6142:791–798

23. Jaszczak PET/SPECT Phantom. http://www.coneinstruments.com
24. Data Spectrum Corporation NEMA IEC Body Phantom SetTM. http://www.spect.com
25. Hoffman EJ, Cutler PD, Digby WM, Mazziotta JC (1990) 3-D phantom to simulate cerebral blood flow and metabolic images for PET. IEEE Trans Nucl Sci 37:616–620
26. DiFilippo FP, Price JP, Kelsch DN, Muzic RF Jr (2004) Porous phantom for PET and SPECT performance evaluation and quality assurance. Med Phys 31:1183–1194
27. RSD Alderson heart/thorax phantom. http://www.rsdphantoms.com
28. Siemens Medical Solutions 2005 Symbia TruePoint SPECT/CT system specifications. http://www.medical.siemens.com
29. Söderberg M, Engeland U, Mattsson S, Ebel G, Leide-Svegborn S (2011) Initial test of a new phantom for investigation of spatial resolution, partial volume effect and detectability in nuclear medicine tomography. Proc J Phys Conf Ser 317:012017
30. Skretting A (2009) 'Intensity diffusion' is a better description than 'partial volume effect'. Eur J Nucl Med Mol Imaging 36:536–537
31. Hoffman EJ, Huang SC, Phelps ME (1979) Quantitation in positron emission computed tomography: 1. Effect of object size. J Comput Assist Tomogr 3:299–308

Chapter 11
Image Magnification with PET Insert Probe

Andrej Studen

11.1 Introduction

This chapter will describe the development of a PET probe based on solid-state detectors. The PET probe concept combines the high spatial resolution of dedicated PET imaging devices with the large field of view of the standard PET scanners. This is achieved by placing an additional movable detector within the standard PET ring and combining standard events with those captured by the probe. The probe must exhibit excellent spatial resolution and be of compact, sturdy design, making solid-state detectors a compelling choice over scintillator-based materials. Although silicon may not be perfectly suited in terms of stopping power and timing resolution, its other characteristics make it a compelling choice for the probe construction. This chapter will give an overview on probe construction, characterization, and performance as well as outline the limiting aspects of the described approach.

11.2 The PET Probe Principle

The landscape of the present PET imaging is the following: On one hand, there are large (and expensive) whole-body PET scanners with, although highly improved in the last years, still somewhat limited spatial resolution [1, 2] and, on the other hand, devices tailored for specific applications (head imaging [3], small animal imaging [4]) with excellent spatial resolution. The question is whether it is possible to combine the best of both worlds, that is, the large field of view with the excellent spatial resolution in the targeted area.

A. Studen (✉)
Jožef Stefan Institute, Jamova 39, 1000 Ljubljana, Slovenia
e-mail: Andrej.studen@ijs.si

A. Giussani and C. Hoeschen (eds.), *Imaging in Nuclear Medicine*,
DOI 10.1007/978-3-642-31415-5_11,

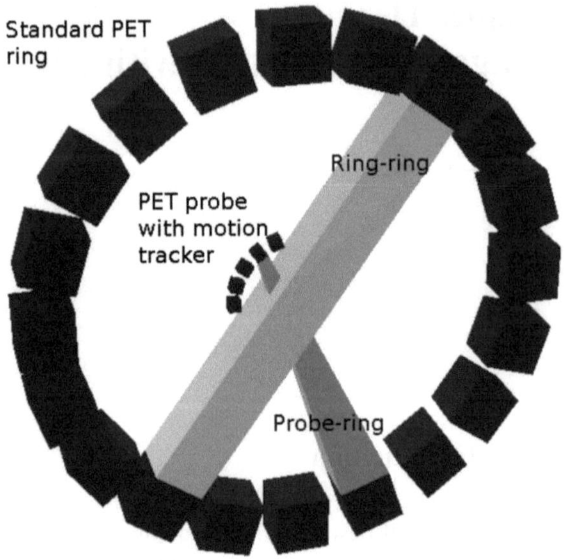

Fig. 11.1 Illustration of the PET probe principle

To achieve that goal, the PET probe concept combines a standard whole-body imager with an additional detector, the probe, which provides additional spatial resolution in selected areas. A possible layout is depicted in Fig. 11.1. As described in Chaps. 3 and 4, the standard PET detector is laid out in a ring to capture both photons generated in the positron annihilation. The positron itself is generated during disintegration of the nuclide embedded in the radiopharmaceutical, with annihilation occurring in the vicinity of the disintegrated nuclide. In Chap. 3, the line of response was defined as that line connecting the interactions of both photons as they are detected in two detector elements on opposite sides of the ring. Since the photons are emitted in nearly opposite directions, the line of response will contain the position of positron annihilation somewhere along its length. A standard interaction, labeled a ring–ring event for categorization purposes, will thus have two photons interacting in ring elements on opposite sides of the ring. The large size of the ring detector cubes shown in the figure serves to indicate the ring's limited spatial resolution, leading to a thick line of response and a large uncertainty in determination of the source position.

The probe provides events of a different type, where one of the two annihilation photons interacts in the additional detector. The excellent spatial resolution of the probe is indicated in the figure by the smaller size of the detector elements. This type of events, categorized as probe–ring events, will define a thinner line of response and consequently delimit a much smaller volume of possible source positions, resulting in an improved inherent spatial resolution.

The idea of combining events from multiple data sets with different resolutions was suggested first by Clinthorne et al. [5]. Their approach was to boost a standard PET ring with an internal high-resolution ring with limited efficiency. The result of their study was quite surprising: Even a relatively small portion of events with high

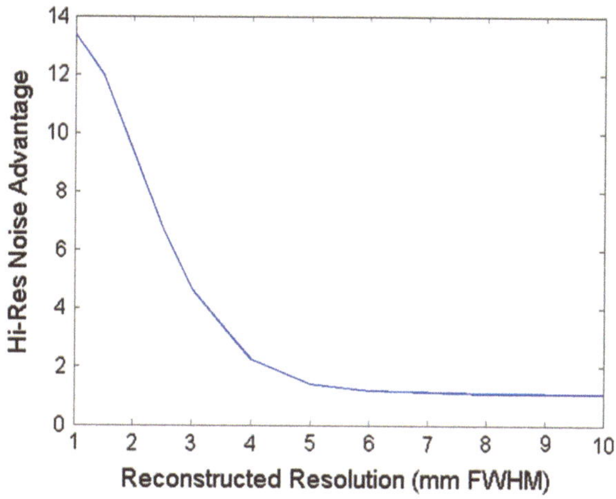

Fig. 11.2 Ratio of image variance for a ring–ring data set versus a data set complemented with 13 % of high-resolution probe–ring data. Simulation assumed a double-ring arrangement, 4-mm FWHM ring, and 1-mm probe resolution

spatial resolution dramatically improves the imaging performance, especially when small details are sought in the image. The imaging performance was quantified via a version of Cramer–Rao bound [6], a curve that relates resolution and variance (or equivalently, contrast and resolution) of a particular realization of reconstructed image. The bound shows a device-characteristic trade-off between the image resolution and the variance and offers a way to compare two independent systems. The results are quoted as a *noise advantage*, expressed as the ratio of the standard deviations in images from the two systems, both reconstructed at the same target resolution.

Figure 11.2 shows the noise advantage of a standard PET ring supplemented by a full internal ring providing an additional 13 % of high-resolution events compared to a stand-alone external ring only. The example shown in the figure refers to a disk of a homogeneous activity and resolutions of standard and probe–ring events assumed to be 4 mm FWHM and 1 mm FWHM, respectively.

There is little improvement in adding high-resolution data while spatial reconstruction remains above to the system resolution of the external ring. Below that margin, the noise advantage starts to rise, reaching a value of 10 at 2-mm resolution. Since for Poisson governed systems the standard deviation is proportional to the square root of the event count, a ratio of 10 in standard deviations translates into 100 times less events required in a mixed system to reach comparable image quality. The anticipated performance of a probe-equipped system will hence rely on the spatial resolution of collected events.

Figure 11.3 shows a probe–ring event with those parameters that are relevant for computing the spatial resolution of such events. The resolution is calculated in the direction perpendicular to the line of response as there is little information on position along the line of response from a single interaction (see also Chap. 3). In that direction the contribution to spatial resolution due to limited resolution of the detector elements will be

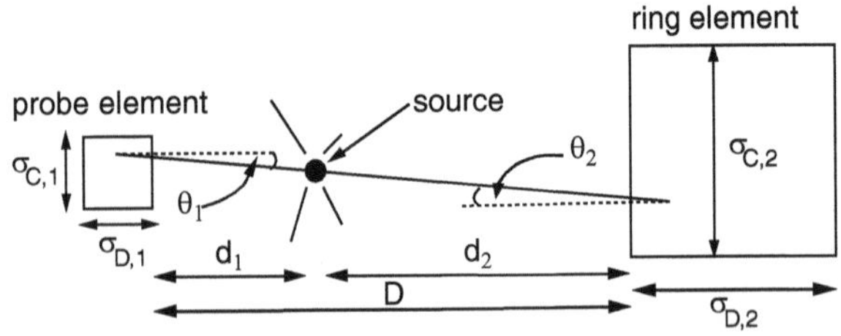

Fig. 11.3 Schematic drawing of the probe and ring detector element and parameters, relevant for computing resolution of probe–ring events

$$\sigma_{\text{tr}}^2 = ((1-\alpha)^2 \cdot (\sigma_{D,1}^2 \sin^2\theta_1 + \sigma_{C,1}^2 \cos^2\theta_1) + \alpha^2 \cdot (\sigma_{D,2}^2 \sin^2\theta_2 + \sigma_{C,2}^2 \cos^2\theta_2)) \tag{11.1}$$

where (see also figure) D is the separation of the sensors, θ_1 and θ_2 are the impact angles, σ_D and σ_C are resolutions of respective detector elements in circumferential and longitudinal (or in depth) direction, and α measures the relative distance d_1/D of annihilation position along the line of response.

For a perpendicular impact, the expression is simplified to α^2-weighted sum of circumferential resolution of both detectors. The resulting curve is drawn in Fig. 11.4 for three conceivable probe intrinsic resolutions. The variance has a broad minimum, and annihilations as far as 10 cm from the probe can still be resolved with spatial resolution nearly identical to the intrinsic spatial resolution of the probe. The effects of annihilation acolinearity have to be added to the above expression, noting however that its spatial resolution also depends on distance of detectors from the annihilation point:

$$\sigma_{\text{AC}} = \frac{d_1 \cdot (D - d_1)}{D} \cdot 0.0088 \tag{11.2}$$

For events close to the probe (in its "focus"), this contribution tends to be quite small (0.7 mm at 10 cm) and can be safely ignored compared to the geometrical resolution given in Fig. 11.4.

From the figure, it can be seen that a probe with a 1-mm inherent resolution will be able to resolve annihilation sources with 1-mm spatial resolution, comparable to high-resolution events from the inner ring of the setup suggested by Clinthorne et al., making all their conclusions valid for the PET probe concept as well.

Since probe–ring events will be mostly collected from a single position of the probe, the probe–ring data set will suffer from limited angle anomalies. However, using the basic image of the external ring and adding the probe–ring data set should circumvent appearance of artifacts in the combined image.

A potential clinical application of the PET probe would be detection and localization of a prostate cancer. The prostate is hard to image as it is located in-between significant amounts of obscuring tissue. For a ring geometry of a PET system, this means that the radiopharmaceutical has to be imaged from a relatively large distances, overlying blurring and noise from the obscuring tissue over the region of interest.

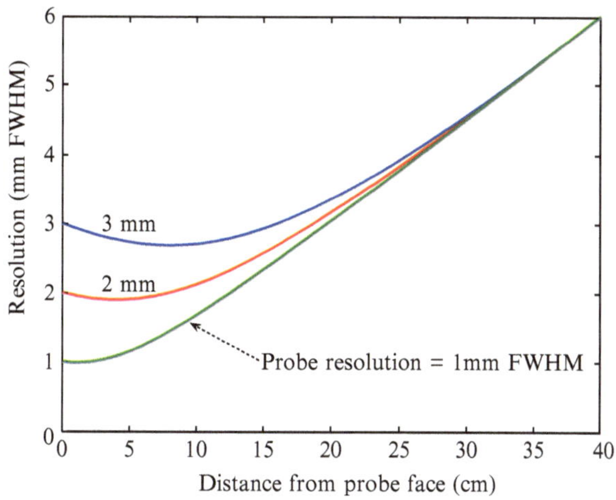

Fig. 11.4 Resolution of the probe–ring event in the direction perpendicular to the line of response. Horizontal axis indicates the distance of the annihilation from the high-resolution detector

Since the outcome of the disease is strongly related to its precise localization (extra-prostatic growth, number of focal points), the current image quality is inadequate for proper staging and treatment planning [7]. One way to overcome present misgivings would be to add a probe to the PET ring and place the probe in the vicinity of the prostate. The probe can explore the proximity of the prostate to the rectal cavity, following examples from transrectal ultrasound [8] and endorectal magnetic resonance imaging [9], or can be a strategically placed external probe. In either case, a near 1-mm resolution of PET-amplified probe in the prostate region could be achieved, allowing for an improved localization and staging of the disease.

11.2.1 The Requirements of the Probe and Selection of Probe Composition

One of the most important features of the detector serving as the probe is its excellent spatial resolution. This should be combined with:

- An excellent timing resolution for proper event matching
- An appropriate energy resolution to be able to recognize events where photons were scattered in the body prior to detection
- Sufficient relative transparency to prevent interference with the external ring while intercepting about 10 % of high-resolution data required to maintain image quality
- Physical compactness and sturdiness for practicality of use

These requirements are hard to satisfy simultaneously. The present detection materials only match some of the requirements while compromise others. Multiple implementations are possible based on either scintillator-based detectors [10] or solid-state detectors [11], which seem to be a compelling choice with practically infinite spatial resolution [12].

The PET probe implementation presented here falls in this latter category. The selected silicon detectors provide unprecedented spatial resolution, by far exceeding requirements of nuclear medicine. In other aspects, the timing resolution is not perfect, but manageable [13], and its low efficiency/stopping power can be compensated by stacking multiple detector layers. The low stopping power can also be seen as an advantage, since external PET ring data will be only slightly modified by the presence of the probe. With robustness of silicon, the practicality issue is also resolved, adding a bonus of the probe being non-perturbed by operation in a magnetic field [14].

While trying to outline general properties of a PET probe, the following sections nevertheless refer to a specific implementation of the probe, developed within the MADEIRA collaborative project.

11.3 The PET Probe Prototype

The PET probe is essentially an addition to an existing PET ring. Consequentially, most efforts were concentrated on the development of the additional detector, the probe, assuming availability of a suitable external ring. In the following discussions, the probe will be assumed to be interfaced to a standard whole-body human PET scanner like those characterized in [1, 2].

11.3.1 Properties of Silicon Detectors

The silicon detectors are standard tools in particle physics. They are used to track charged particles close to the interaction point. The silicon detectors intended for the PET probe are a slightly modified version of the originals.

The basic detector layout is illustrated in Fig. 11.5. The detector is a reversely operated diode with varied doping concentration. The central part (bulk) of the detector is made from monocrystalline silicon of high purity and consequently resistivity in excess of 30 kΩ cm. A thin (a few μm) top p-layer with strong doping concentration ($10^{24} \times m^{-3}$) forms a p–n junction with the bulk, and a similar n-type layer at the bottom provides low-resistivity interface to the backplane electrode. The top p-layer is segmented to independent pads. The size of the pads of detectors for PET imaging is set at around 1 mm, small enough to provide sufficient resolution and large enough to guarantee that the generated charge is collected in a single pad for each event.

The detector thickness is modified from the standard particle detectors. A thickness between 0.5 and 1 mm is a good compromise between efficiency, timing resolution, channel count, and limitation of planar detector processing.

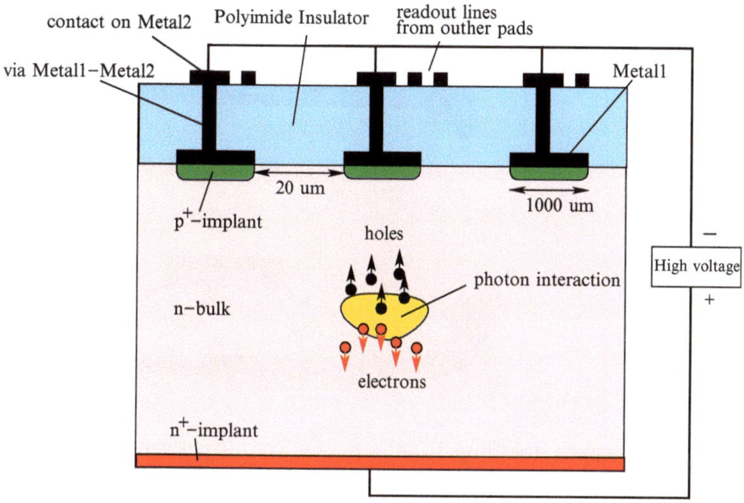

Fig. 11.5 Schematic illustration of silicon as position-sensitive photon detector with sketched device geometry, application of high voltage, and carrier movement for a photon interaction

With attenuation coefficient of $\mu = 0.2\ \text{cm}^{-1}$ for 511-keV photons, a 1-mm-thick detector stops about 2 % of incoming photons. With standard PET detector elements being nearly 100 % efficient, five or more detector layers must be used to reach the desired 10 % ratio of high-resolution events in the sample.

The detector is reversely biased to voltages exceeding the full depletion voltage, depleting the detector of thermally generated free carriers. An interaction in the detector produces countable pairs of free electrons and holes, the number of which is proportional to the energy deposited by the interacting particle. The generated charge is driven by the electric field to induce signal on electrodes attached to the pads.

The signals collected on the electrodes (10^5 electrons) are relatively small compared to the signals in the standard PMT-based PET detectors (10^9 electrons), which makes silicon signals noise prone. This requires dedicated low-noise readout electronics placed very close to the sensors.

11.3.2 The PET Probe Module

The MADEIRA PET probe was constructed of several identical, largely autonomous modules.

The MADEIRA probe module, photographed in Fig. 11.6, is composed of a pair of silicon detectors, each measuring $40 \times 26\ \text{mm}^2$ and 1 mm thick. Only the top detector is seen in the photograph, the bottom one is exactly behind it, with an approximate spacing of 0.8 mm between the detectors. This spacing is to be small to maximize content of active material in the probe. Each detector is segmented to

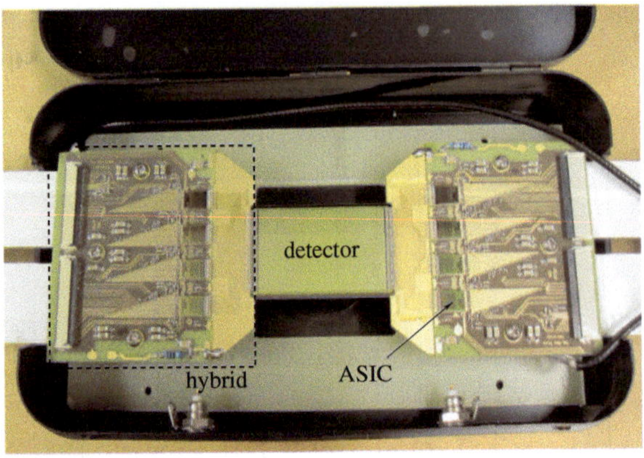

Fig. 11.6 Photograph of a silicon module used as a building block of the silicon PET probe

$1 \times 1\text{-mm}^2$ square pads, totalling 1,040 sensitive voxels per sensor, each of 1-mm^3 volume and independent from one another.

The detectors are attached to the hybrids, printed circuit boards (PCBs) that provide mechanical support for the detectors and host the first stage of electronic processing. We used VATAGP7, application-specific integrated circuits (ASICs) made by IDEAS [15] for front-end electronics. Each VATAGP7 chip hosts 128 identical parallel processing channels, each channel composed of a low-noise charge-sensitive preamplifier, whose output is split to a pair of electrically separated CR-RC shapers with different shaping times. The fast shaper with a shaping time of 150 ns serves as an input to a leading edge discriminator (one per channel) with a common chip threshold. A fine adjustment of thresholds for each channel is available through onboard digital-to-analog converters (DACs). A logical OR function of all trigger signals from all channels within a chip gives the chip trigger. The slow shaper with a shaping time of 500 ns provides an analog signal corresponding to the number of electron holes collected in this pad or equivalently to the deposited energy. The analog signals are placed in an on-chip shift register. The shift register can be read in a serial mode, when signals collected in all channels on the chip are read sequentially by clocking through the register, or in a sparse readout mode, when only energy of channels that gave a trigger are read from the shift register.

In total, 16 ASICs are required to read the 2,080 pads of both silicon sensors. These are split among four identical PCB boards. A flat cable connector connects the ASICs to the downstream electronics.

11.3.3 The Data Acquisition System

The task of the data acquisition system (DAQ) is to manage the data coming from the probe and interface it to the external ring. On top of that, the DAQ should provide a way to set the operating conditions of the probe. Because of the modular approach, the DAQ should be scalable as well.

Fig. 11.7 Photograph of a MADDAQ board attached to a test hybrid

For the prototype, a dedicated DAQ system was constructed under a name of MADDAQ [16], depicted in Fig. 11.7. Each MADDAQ board serves one probe module and interfaces directly into a personal computer (PC) via an Ethernet cable. The Ethernet protocol offers an attractive way of multiplying the MADDAQ boards with minimum hardware overhead. Each board is powered with laboratory voltage supplies (+3.3 V and −3.3 V required) and also powers the remote ASICs. The reverse bias voltage of 200 V for the sensors bypasses the MADDAQ and is provided directly to the modules.

The MADDAQ board was built around a field programmable gate array (FPGA) called Spartan-3 from Xilinx [17]. The FPGA and associated circuitry provide the following features:

- A trigger logic implementation. The trigger from the external ring and the triggers from the VATAGP7 chips are combined, and coherence is tested, via a standard coincidence time window or through a more complex trigger algorithm
- Event readout block, sequencing VATAGP7 readout, analog-to-digital conversion of channel energies, and packaging of the event to data words
- Slow control of the data acquisition, including setting of the thresholds for VATAGP7 discriminators, utility to program VATAGP7 register controlling chip operation and provide settings for the low-noise preamplifiers
- Communication utilities that regulate transfer of collected data to the PC and interpret the slow control commands coming from the control PC

11.4 Module Characterization

The modules were characterized with reference to parameters relevant for probe performance. Here the expected efficiency, the energy resolution of the modules, and their timing resolution will be reported.

Table 11.1 Frequencies of event types in a PET probe configuration for a single 1-mm layer of the MADEIRA PET probe detector and a standard-type external ring detector

Event type	No. of interactions per 1,000 events
Total probe interactions (including singles)	4.5
Coincidences	0.81
Coincidences of scattered photons	0.75
Clean coincidences	0.06
Ring–ring clean coincidences	93.9

Photon pairs with opposite directions were generated randomly within an ellipsoidal cylinder filled with water, measuring 60 cm axially and between 10 cm and 20 cm trans-axially. The ring and the probe were centered on the object. Events that scattered prior to interaction were tagged and counted separately. The number of detections per 1,000 of generated pairs is shown

11.4.1 The Expected Efficiency of the Probe

The efficiency of the probe was estimated through simulation [18]. We wanted to determine approximate relative number of probe–ring events versus ring–ring events, grouped by location of annihilation, for a very general probe position and a very general imaging setting. Geant4 [19], a standard software simulation tool, was used for tracking photons through the object and the detection system.

For the generalized object, an ellipsoidal cylinder filled with water was used with approximate human torso dimensions. For the generalized probe detector, a standard imaging ring was supplemented by a single layer of the probe with the dimensions of the silicon detectors, with the probe laying facedown 1 cm from the surface of the object. Pairs of photons headed in opposite directions were randomly generated within the object, and the detected events were sorted by event type, separating clean events from those that scattered prior to interaction. These results are listed in Table 11.1. This points to a relative inefficiency of the probe–ring events compared to the ring–ring events when a uniform background is imaged.

To get a more detailed picture of where the detected events are coming from, the object was coarsely divided to large voxels. Only clean detections were counted, assuming a viable scattering event recognition, and their ratio was plotted versus the voxel location. This is illustrated in Fig. 11.8 which shows a transverse and an axial slice through the object at its center. Clearly, the probe is a focusing device with most interactions coming from its immediate vicinity where the relative efficiency increases to 1 % of probe–ring interactions in the data set.

Several important conclusions follow:

- A set of five to ten detectors will provide sufficient sampling to introduce the drastic image quality improvements suggested by Clinthorne et al. portrayed in Fig. 11.2.
- The sensitivity of the probe to the nearby annihilations only will cause most events to share the excellent spatial resolution of the probe projected in Fig. 11.4.
- Probe will efficiently reject all out of focus events by 0.01 % overall efficiency compared to 1 % "in focus" efficiency.

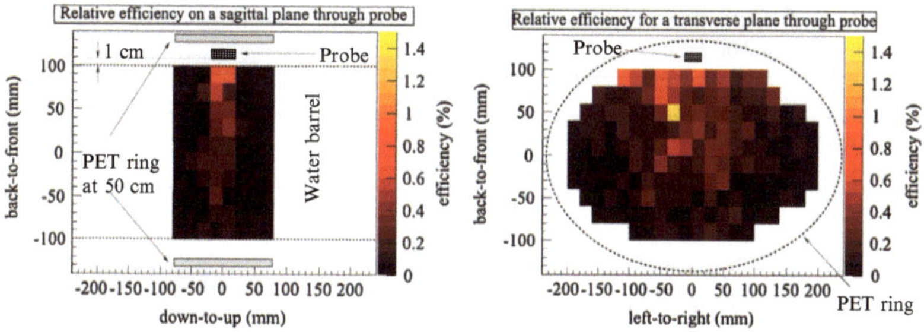

Fig. 11.8 Relative efficiency of probe–ring event type versus ring–ring event types from simulation. Two cross sections through simulated water phantom are given, trans-axial on the left and axial (side) view on the right. The relative percentage is color coded according to colormap shown. Drawings indicate approximate positions of the probe and the external ring. Notice that the ring is drawn out of scale. The probe dimensions were equal to the size of the silicon sensors, $2 \times 4\ \text{cm}^2$, with long edge oriented in the left-to-right direction

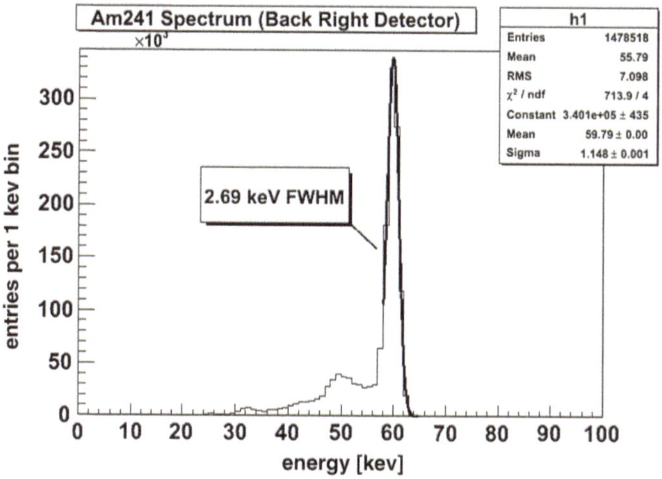

Fig. 11.9 Spectrum of gammas from ^{241}Am as recorded in the detector module

11.4.2 *The Energy Resolution*

Strictly speaking, the energy resolution has no bearing on the probe performance. However, energy resolution is a strong indicator of the noise contamination of the signal readout, and a proper energy resolution indicates proper operation of the ASIC. Implicitly, however, the energy resolution has a strong bearing on event classification and timing resolution. The energy resolution is measured by exposing the sensor to radiation from a calibration source, ^{241}Am with a clean γ-ray at 59.5 keV. Figure 11.9 shows a spectrum obtained with a MADEIRA probe module. The width of the line corresponding to photoelectric effect of incoming photons is 2.7 keV FWHM.

Table 11.2 Effectiveness of classification based on energy recorded in a probe (silicon) sensor for an energy cut at 100 keV

True event type	All	Assumed clean	Assumed scattered
Scattered prior to interaction	27.6	1.7	25.9
Clean	12.4	6.7	5.7

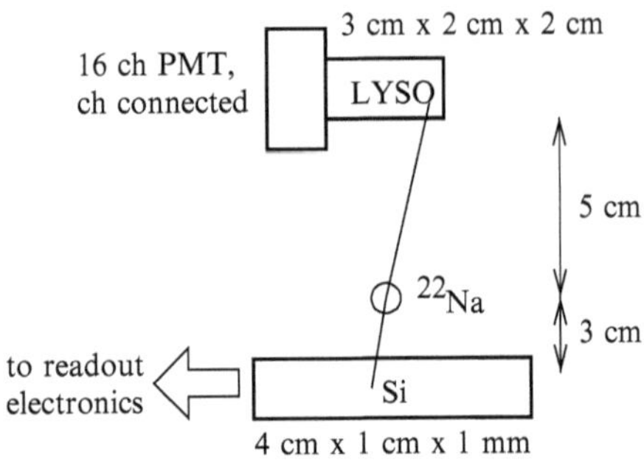

Fig. 11.10 Sketch of the setup used to verify the timing performance of the probe module

During the simulation of probe efficiency, it was possible to correlate energy recorded in silicon sensor to photon history, which is scattering prior to detection. Since photons lose energy in interactions, the high-energy part of the spectrum is much less contaminated with scattering than the lower part. Since the energy resolution is excellent, a very precise cut on energy could be made to optimize energy filtering. Table 11.2 shows effectiveness of a cut at 100 keV both in terms of properly recognized scattered photons as well as in terms of misclassified events.

11.4.3 The Timing Resolution

Timing resolution is very important for sensor synchronization, crucial for event matching. Here results obtained with a MADEIRA module prototype as described in [13] will be presented. A simple setup portrayed in Fig. 11.10 was used to measure the effect. The silicon detector under test was a simplified version of the module presented above, with a smaller, $1 \times 4\text{-cm}^2$ sensor divided into pads with 1.4-mm side, requiring a single hybrid readout with two VATAGP7 only. The detectors were however cut from the same silicon wafers as the detectors for the probe and the ASICs and the hybrids were identical to those of the MADEIRA modules. As a timing reference detector, a LYSO-photomultiplier-tube (PMT) assembly with a timing resolution of below 1 ns FWHM was used. Detectors were facing each other, and a positron source was inserted in between. The emitted positrons promptly annihilated to photon pairs, each striking one of the sensors. Synchrony of detector response was measured by recording the delay between the

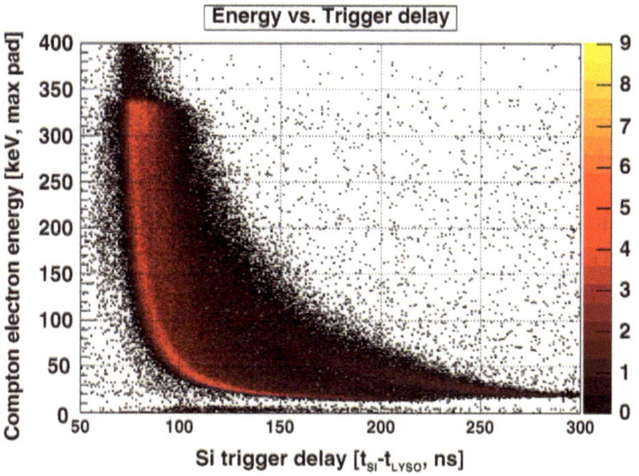

Fig. 11.11 Timing characterization events: distribution of events according to the energy measured in silicon and delay of silicon trigger after trigger from the LYSO timing reference

trigger from the LYSO-PMT and the module trigger. As the trigger for silicon is generated by a leading edge discriminator implemented in VATAGP7, the trigger time will be correlated to the signal amplitude, a phenomenon referred to as the time walk. To record the time walk, the signal amplitude which corresponds to the measured energy deposit in the detector was recorded with each event. Since Compton scattering dominates at the interactions of 511-keV photons in silicon, the recorded energy will be distributed according to the continuous, Klein-Nishina energy distribution.

The correlation of energy and delay is shown in Fig. 11.11. The color drawn at a given location in the plot represents relative frequency of events with a corresponding combination of delay and energy measurements, with black indicating a low and yellow indicating a high event count. The red and yellow areas trace out the characteristic time-walk correlation, while the steepness of the drop from the ridge indicates the quality of the timing resolution. The figure indicates resolutions in the range of a few tens of ns, dominated by position-related jitter of the trigger delay. Namely, the electron–hole cloud excited by the photon-excited electron extends only a small distance (up to 300 μm) from the photon impact. During charge collection, the speed of the carriers is proportional to the electric field, which is linear and strongest at the p–n junction at the top of the sensor and smallest near the backplane. Should the cloud creation occur in the low-field region, the initial velocity is small, resulting slope of the signal is small, delaying the threshold transition. In the opposite case of carriers generated in the high-field region, the signal grows quickly causing a prompt threshold transition. The effect is subdued by increasing the reverse voltage in the detector, urging us to operate at the highest biases tolerated by the particular detector design.

Since the position-related jitter is highly asymmetric, it is difficult to classify the resolution in terms of standard parameters, for example, a FWHM. Rather, the negative impact on the data classification was studied through the coincidence window algorithm. This standard method assumes that triggers arriving within a time window of duration t_W are caused by the same positron annihilation. Prior to coincidence logic, the triggers can be compensated for mismatches in cable lengths and delays of the individual logic circuits.

For a *true event*, both photons from a specific annihilation must first reach their respective detectors. Even for a valid hit in a silicon detector, there will be a limited number of events where the second photon will strike a ring element as well. This probability is clearly a combination of the object properties and the geometry of the readout. For the sake of argument, a number from simulation given in Table 11.1 can be used where on average 0.19 coincidences were recorded per a valid probe interaction, with the other photons scattering out of the field of view of the external ring. This number will be further degraded should the distribution of the trigger delays for valid events extend beyond the desired time window t_W. In that case, only a portion of events with trigger times within t_W will contribute to the probability of a true event. In this way, a specific timing distribution is expressed as a probability and can be compared to the probability of a random event.

A *random event* will be admitted when the ring and the probe event were caused by two separate positrons emitted at different times and different locations but were recorded within the time window of duration t_W. The probability of that happening is $P_R = 1 - \exp(-A \cdot \varepsilon \cdot t_W)$ where A is the activity viewed by the external ring and ε is the probability that a positron emitted anywhere will give at least one interaction in the ring. The parameter ε will depend on combination of detector geometry, object geometry, and activity in the object. For the sake of simplicity, the value of ε obtained in the simulation above for a generalized object will be used.

Figure 11.12 shows a comparison of both random and true event probabilities for a set of given activities and parameters obtained from the simulation. Two sets of probe performances are drawn—expected performance of the current 1-mm-thick probe and of a probe made of half as thin detectors (0.5 mm). The first conclusion is that although extending time window improves chances of capturing a true event, the contamination with random coincidences overwhelms the benefits. The second conclusion would be that thinner detectors can be used in a much harsher activity environment.

11.4.4 Spatial Resolution

The spatial resolution of silicon detectors is determined by the pitch of the segmentation. In the MADEIRA probe development, two sets of detectors were implemented, identical (same material, same wafers, same metallization thickness) save for the size of the pads. The first probe generation has square pads with a side of 1.4 mm, while the second generation has a slightly smaller side of 1 mm.

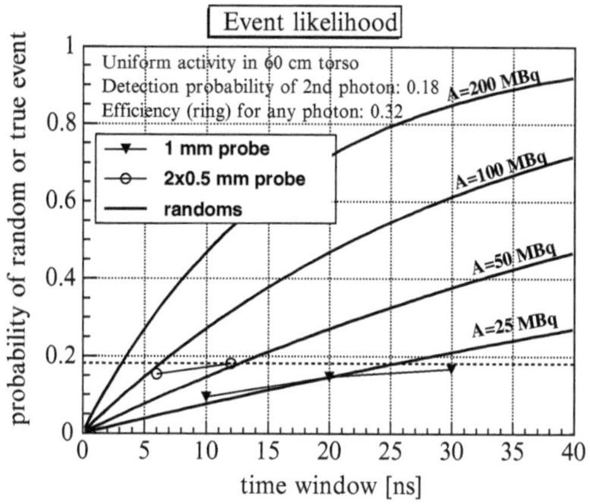

Fig. 11.12 Expected probabilities that random or true coincidences will be present in the data set versus a selected time window duration. Probabilities of random coincidences are for listed total activities of ^{18}F-like emitter in the simulated water phantom. The probabilities for true coincidences for the existing probe (*inverted triangles*) and for a potential high-activity adjusted probe composed of twice the detector number with half (0.5 mm) the thickness (*open circles*)

11.5 Performance of the Probe

The probe was interfaced to a laboratory PET setup constructed at the University of Michigan. The test setup was built specifically to tests new PET imaging devices and provided simple interface to the probe electronics.

11.5.1 PET Test Setup

The PET test setup (photographed with the probes in Fig. 11.13) was built from standard PET BGO modules from a CTI 931 PET scanner. Each module is composed of a 53.4 × 50-mm^2 BGO crystal, segmented to 4 × 8 cells. The segmentation is not complete as to allow light sharing among four PM tubes. The output is processed by a custom-made electronics board, which amplifies the signals of all four tubes and provides a leading edge trigger signal. The trigger signal is well aligned, and a timing resolution of 10 ns FWHM was measured. The modules were placed in two arcs with a radius of 50 cm, each arc containing 12 modules and providing 67.5° coverage with respect to the circle center. The modules were oriented with the fine segmentation circumferentially. Signals of all tubes were fed into a peak sensing ADC circuit (VME V785 ADC [20]) where the peak value was digitized and stored. For spatial and energy calibration, the setup was exposed to flood illumination from a ^{22}Na source, which allowed for clear separation of signals and expected energy resolution of about 23 % at the 511 keV photo-peak.

Fig. 11.13 A photograph of the test setup for probe characterization; *left*: a full view, including the (partial) external ring, and *right*: a zoom on the probe arrangement

This configuration allowed multiple detector arrangements. The photograph shows a placement of two MADEIRA probe modules of the first generation in an edge-on orientation. In this orientation, the sensors were aligned to a single imaging plane, parallel to the optical table, for a maximum sensor efficiency. The detectors were placed 66 mm from the center of the field of view behind a planar lead collimator that constrained the acceptance of the external ring to the plane of the probe detectors. The object itself was placed within the collimator and on a rotary table to allow object rotation within the full circle providing the full tomographic data sample.

The probe readout was a simplified version of the MADDAQ called VMEDAQ, which uses the same VME back end for sensor control and data transfer as the PET setup. The trigger implementation, normally residing within the MADDAQ FPGA, was implemented in a general purpose VME board (V1495 FPGA [20]).

11.5.2 Spatial Resolution Studies

The event system resolution was determined using point-like sources (1-mm diameter) of ^{22}Na placed on the rotating table. A full rotation with 6° rotation steps and 2 min dwell time at each position was performed. Such a scan yielded 300,000 probe–ring events, which were reconstructed on a $0.2 \times 0.2\text{-mm}^2$ imaging grid. A maximum likelihood expectation maximization algorithm with no additional blurring was used to reconstruct images.

Figure 11.14 shows a reconstruction of a combined data set where the source was shifted in 2-mm steps from the center of the field of view. The converged images show a resolution between 1.4 and 1.5 mm which is consistent with a pad size of 1.4 mm, used in this test. Looking back at Fig. 11.4, a system resolution of the probe–ring events does indeed match the segmentation of the probe at distances between 6 and 10 cm, roughly corresponding to the broad resolution minimum depicted in Fig. 11.4.

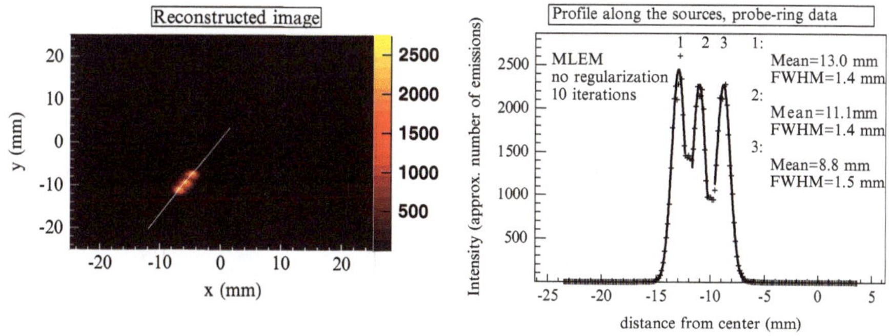

Fig. 11.14 Image reconstruction of a set of three ^{22}Na point-like (1-mm diameter) source. *Left*: reconstructed image. *Right*: profile through sources along the *white line* indicated on the reconstructed image. Gaussian fits to peaks labeled numerically from left to right yield indicated means and resolution

11.5.3 Small Field of View Phantom Studies

The probe principle in edge-on geometry was also verified. A high-resolution Jaszczak phantom (hole diameters of 1.2, 1.6, 2.4, 3.2, 4.0, and 4.8 mm) filled with a ^{18}F-FDG solution was used. In the current setup, two runs were necessary. First, the base image was recorded with ring–ring events collected for 20 rotations of the phantom with a dwell time of 15 s at each of the 6° angular rotation steps. For an initial specific activity of 40 MBq/mL, a total of ten million of ring–ring events were recorded of which five million passed selection criteria (energy window of 300–700 keV, no inter-crystal scattering, no multiple events) and contributed to the image. In the second run, the phantom was nearly stationary, with a total angular range of 12° only serving to virtually extend the size of the probe by about 25 % and provide fine event sampling. The initial specific activity and the total data collection time were equal for both data sets, making one million of probe–ring events a matched high-resolution data sample.

Figure 11.15 shows the measurements setup and resulting images. Due to its position in the setup (Fig. 11.15a), the probe collected lines of response predominantly oriented along the *Y* axis. As a result of that, the hot spots are blurred to lines along the unsampled *Y* direction (Fig. 11.15c). Only where coarse resolution of the base image is sufficient, for example, in 4.8-mm spots, fidelity with the original image is maintained in that direction as well. The probe performed much better in direction perpendicular to the collected lines of response. The white line in the reconstructions shows where profiles along *X* were cut through both the base image in Fig. 11.15b and the augmented image in Fig. 11.15c. The cut was made at 15 mm offset from the phantom center towards the probe where both probe sensitivity and resolution are substantial. While the base image profile shows little variation stemming from the 2.4- and 1.6-mm-diameter hot spots, the augmented image clearly resolves the *Y*-blurred hot spots to individual peaks. In this phantom, the spots were separated by twice the spot diameter, and their hexagonal packing makes

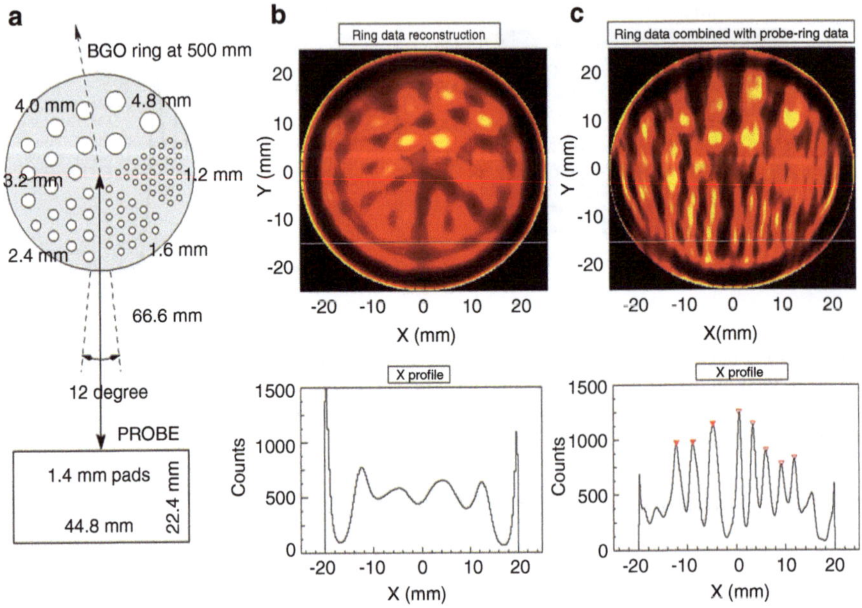

Fig. 11.15 Image reconstruction of ^{18}F-filled Jaszczak phantom recorded by the probe characterization setup. (**a**) Schematic of the phantom, its angular shift during the data collection and the position of the probe. (**b**) Data reconstructed from ring–ring data, with profile along the line marked in the reconstructed image. (**c**) Reconstruction of the combined data set with profile along the line marked in the reconstruction image. Data collection time was ~5 h, EM-ML reconstruction with 400 iterations was used

rows separated by the height of the corresponding equilateral triangle. In 2.4-mm spot diameter region, the actual separation is 4.2 mm, and the measured shifts between peaks marked with solid inverted triangles in (Fig. 11.15c) are 4.0 and 3.4 mm as we move out from the center. In the 1.6 mm, the expected spacing is 2.8 mm, and the measured offsets between peaks marked with open inverted triangles yield distances of 2.9, 2.6, 3.1, and 2.6 mm as we move out from the center. It can be concluded that in both regions the probe provides both image resolution surpassing 2 mm and excellent image fidelity. For reconstruction, EM ML algorithm was used. In mixed data case, each data set was treated independently with sensitivity scaled to reflect the event count in each of the event categories, and 400 iterations were used.

11.6 Conclusion

A device operable as a PET probe was successfully designed and instrumented. The MADEIRA PET probe is based on high-resistivity silicon sensors and provides the required high spatial-resolution data for moderate activity environments. The probe was interfaced to a test PET ring setup, and the initial probe characterization was performed. A simplified geometry was used where the events were mechanically collimated to a single imaging plane. In this setup the systematic resolution of the probe–ring events matched the spatial resolution of the probe, confirming the predicted behavior. The image reconstructed from ring–ring events combined with probe–ring data clearly resolves 1.6- and 2.4-mm holes in a high-resolution Jaszczak phantom, which were completely obscured in the bare ring–ring data, confirming the PET probe principle in a simplified planar geometry. More elaborate imaging examples are being performed and will be reported elsewhere. The data presented shows that the probe can indeed combine large field of view and high spatial resolution in a single device, a fact which might prove important in the future of PET imaging.

References

1. Surti S, Kuhn A, Werner ME, Perkins AE, Kolthammer J, Karp JS (2007) Performance of Philips Gemini TF PET/CT scanner with special consideration for its time-of-flight imaging capabilities. J Nucl Med 48(3):471–480
2. Jakoby BW, Bercier Y, Conti M, Casey M, Gremillion T, Hayden C, Bendriem B, Townsend DW (2008) Performance investigation of a time-of-flight PET/CT scanner. In: IEEE nuclear science symposium conference record, Oct 2008, pp 3738–3743
3. Wienhard K, Schmand M, Casey ME, Baker K, Bao J, Eriksson L, Jones WF, Knoess C, Lenox M, Lercher M, Luk P, Michel C, Reed JH, Richerzhagen N, Treffert J, Vollmar S, Young JW, Heiss WD, Nutt R (2002) The ECAT HRRT: performance and first clinical application of the new high resolution research tomograph. IEEE Trans Nucl Sci 49(1): 104–110
4. Kim JS, Lee JS, Im KC, Kim SJ, Kim S-Y, Lee DS, Moon DH (2007) Performance measurement of the microPET focus 120 scanner. J Nucl Med 48(9):1527–1535
5. Clinthorne NH, Park S, Wilderman SJ, Rogers WL (2001) High resolution PET detector. J Nucl Med 42(5):383
6. Meng LJ, Clinthorne NH (2004) A modified uniform cramer-rao bound for multiple pinhole aperture design. IEEE Trans Med Imaging 23(7):896–902
7. Oehr P, Bouchelouche K (2007) Imaging of prostate cancer. Curr Opin Oncol 19:259–264
8. Ragde H, Kenny GM, Murphy GP, Landin K (1997) Transrectal ultrasound microbubble contrast angiography of the prostate. Prostate 32(4):279–283
9. Metzger GJ, van de Moortele P-F, Akgun C, Snyder CJ, Moeller S, Strupp J, Andersen P, Shrivastava D, Vaughan T, Ugurbil K, Adriany G (2010) Performance of external and internal coil configurations for prostate investigations at 7 t. Magn Reson Med 64(6):1625–1639
10. Tai YC, Wu H, Pal D, O'Sullivan J (2008) Virtual-pinhole PET. J Nucl Med 49(3):471–479
11. Yin Y, Komarov S, Wu H, Song TY, Li Q, Garson A, Lee K, Simburger P, Dowkontt P, Krawczynski H, Tai YC (2009). Characterization of highly pixelated czt detectors for sub-millimeter pet imaging. In: IEEE nuclear science symposium conference record (NSS/MIC), pp 2411–2414
12. Straver J, Toker O, Weilhammer P, Colledani C, Dulinski W, Turchetta R, Bosisio L (1994) One micron resolution with silicon strip detectors. Nucl Instrum Methods Phys Res A 348 (2–3):485–490
13. Studen A, Burdette D, Chesi E, Cindro V, Clinthorne NH, Cochran E, Grošičar B, Kagan H, Lacasta C, Linhart V, Mikuž M, Stankova V, Weilhammer P, Žontar D (2010) Timing performance of the silicon PET insert probe. Radiat Prot Dosimetry 139:199–203

14. Burdette D, Albani D, Chesi E, Clinthorne NH, Cochran E, Honscheid K, Huh SS, Kagan H, Knopp M, Lacasta C, Mikuz M, Schmalbrock P, Studen A, Weilhammer P (2009) A device to measure the effects of strong magnetic fields on the image resolution of pet scanners. Nucl Instrum Methods Phys Res A 609(2–3):263–271
15. Gamma Medica-Ideas, Inc. (Norway). http://www.gm-ideas.com. Martin Linges vei 25, Snarøya, POB 1, N-1330 Fornebu, Norway, Tel: +47-6782-7171
16. Stankova V, Lacasta C, Linhart V, Llosa G, Burdette D, Cindro V, Cochran E, Chesi E, Clinthorne NH, Grosicar B, Honscheid K, Huh S, Kagan H, Mikuz M, Solaz C, Studen A, Weilhammer P, Zontar D (2010) An fpga based daq system for the readout of madeira pet probe. In: IEEE nuclear science symposium conference record, 2010. NSS'10
17. Xilinx, Inc. http://www.xilinx.com. 2100 Logic Drive, San Jose, CA 95124-3400
18. Studen A, Chesi E, Cindro V, Clinthorne NH, Cochran E, Grosicar B, Honscheid K, Kagan H, Lacasta C, Llosa G, Linhart V, Mikuz M, Stankova V, Weilhammer P, Zontar D (2010) A silicon pet probe. Nucl Instrum Methods Phys Res A 648(Suppl 1):S255–S258
19. Agostinelli S et al (2003) GEANT4: a simulation toolkit. Nucl Instrum Methods Phys Res A 506:250–303
20. CAEN, S.p.a. http://www.caen.it. Via Vetraia, 11, 55049 Viareggio (LU) Italy. Tel.: +39.0584.388.398

Part IV
Optimization of Nuclear Medical Imaging

Chapter 12
Rules of the Thumb

Christoph Hoeschen

The overarching goal of nuclear medical imaging is to provide a diagnosis for a patient as accurate as possible. The main advantage for nuclear medical imaging compared to other methods is the ability to provide functional information. In particular, the sensitivity is much higher than with any other medical imaging method (see Fig. 12.1). On the other hand, the spatial resolution and temporal resolution of the nuclear medical imaging methods are typically limited compared to other imaging methods. In addition, nuclear medical imaging is always combating the most relevant problem in imaging which is noise.

Although noise is a problem in all medical imaging applications, it is prominent in medical imaging using ionizing radiation due to the fact that from the physics point of view, lower noise can only be achieved by higher doses due to X-rays or radioisotopes, which would result in higher exposures of the patient. Since the amount of applied radioactive material to a human being has to be limited as much as possible, the noise problem is very prominent in nuclear medical applications.

The goal of optimizing the nuclear medical imaging technologies and methods is to provide better diagnostic performance which means a better or more secure performance, which would in the best case result in a possible earlier diagnosis of a disease or even a complete new diagnosis. As a crude approximation, the clinical performance (CP) can be described by the following equation:

$$\mathrm{CP} = \mathrm{HOP} * \mathrm{DQE}_{\mathrm{Pres}} * \mathrm{DQE}_{\mathrm{Detect}} * \mathrm{NEQ}_{\mathrm{Gen}}$$

where HOP means the human observer performance which is obviously only somehow correlated to the image quality, $\mathrm{DQE}_{\mathrm{Pres}}$ is the detective quantum

C. Hoeschen (✉)
Research Unit Medical Radiation Physics and Diagnostics, Helmholtz Zentrum München – German Research Center for Environmental Health, Ingolstädter Landstr. 1, 85764, Neuherberg, Germany
e-mail: christoph.hoeschen@helmholtz-muenchen.de

A. Giussani and C. Hoeschen (eds.), *Imaging in Nuclear Medicine*,
DOI 10.1007/978-3-642-31415-5_12, © Springer-Verlag Berlin Heidelberg 2013

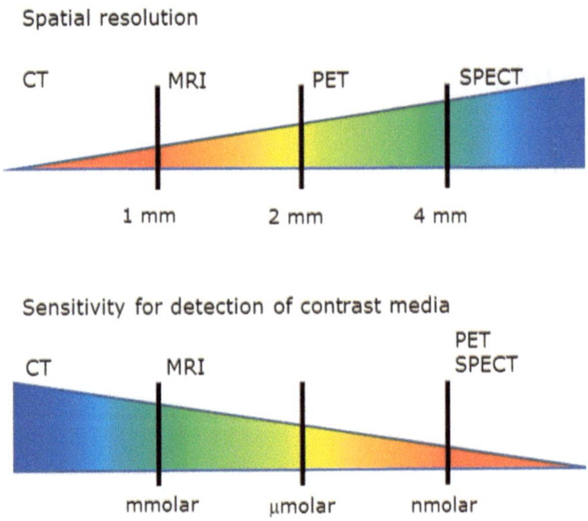

Fig. 12.1 Properties (resolution and sensitivity) of various medical imaging methods

efficiency of the presentation procedure, DQE_{Detect} is the detective quantum efficiency of the detection process and NEQ_{Gen} is the noise equivalent quanta of the signal generation. The latter is influenced by the biokinetics, the dose modeling for each patient, and the time scheme of the imaging process. DQE_{Detect} depends on the detector efficiency, the spatial resolution properties, and the calibration facilities, while the presentation DQE mainly reflects the efficiency of the reconstruction and image processing procedures.

This goal of optimizing the nuclear medical imaging performance in terms of an optimized tool for diagnosis must be correlated with the general principle of optimization of radiation protection of keeping doses "as low as reasonable achievable" [1] as it is valid according to legal requirements for the use of ionizing radiation especially in the context of medical applications.

In respect of this, optimization needs contributions from multiple disciplines:

1. Clinicians providing clear clinical tasks and make a clear indication case
2. Physicians, biologists, and chemists developing sensitive and specific new biomarkers
3. Physicians and biophysicists investigating bio- and pharmacokinetics
4. Medical physicists and engineers providing tools for optimized spatial and or temporal resolution, taking physical and practical limitations into account
5. Mathematicians and software developers contributing algorithms and software packages specifically designed for the development

In the following, the above aspects will be briefly described in its connection with the purpose of optimization and the other components of this optimization process.

12.1 Clinicians Providing Clear Clinical Tasks and Make a Clear Indication Case

As a part of the ALARA principle, it is necessary to avoid all kinds of exposures to human beings which are not justified, e.g., by medical needs. In this regard it is really necessary to identify and describe the intended benefit for the patient and balance it versus the estimated doses and the corresponding risks. For further aspects of this, we refer to the rules of the thumb chapter in the second book of this series [2]. The description of the clinical task has further aspects in the optimization procedure. First of all, depending on the task, the ideal imaging system might be different. In connection with this, the optimal radiopharmaceutical and the optimal dosage have to be determined. This refers to actual available radiopharmaceuticals and data but could be changing according to the optimization processes described in the following. The last aspect about the optimization regarding the description of the clinical task is the choosing of the optimal time point for the investigation. This will depend on the accumulation of the radiopharmaceutical within the lesion of interest compared to the background.

12.2 Physicians, Biologists, and Chemists Developing New Sensitive and Specific New Biomarkers

As indicated already before, optimal biomarkers are needed able to bind to the diseased cells in question (sensitivity) or characterize the activity of a tissue. It should be avoided that the biomarkers also bind to cells characterizing normal tissues and therefore not being specific markers for the pathology. Not all radiopharmaceuticals available today fulfill these conditions. This lack of specificity and/or sensitivity might result in higher exposures necessary to gain a diagnostic image useful for clinical diagnosis. The importance of an optimal sensitivity and specificity as well as the signal-to-noise ratio depends also on the clinical task in question.

If a new biomarker can be found which binds very effectively to the diseased cells in question and is in addition quite specific, it is necessary to find a way to couple this biomarker to a radioactive isotope. Here the first aspect for the optimization is certainly the necessity of a stabile connection between the marker and the radioisotope. In addition, the selected radioisotope has to be a radiation source suitable for imaging (positron emitter, gamma emitter), and the physical half life should fit to the production, transportation, and imaging procedure [3] as well as to the biological medical needs.

12.3 Physicians and Biophysicists Investigating Bio- and Pharmacokinetics

As mentioned in the paragraph before, during the development of radiopharmaceuticals, the fitting of the radioisotope to the biological clinical setting is of great importance. This is connected with the bio- and pharmacokinetics of the biomarker in question. So, it is important to study in detail the time dependence of the distribution of the biomarker in a healthy human being as well as in the pathology. As it could be shown within the MADEIRA project, the different distribution of even quite standard markers in the pathological tissue and the surrounding healthy tissue might result in an optimized time schedule for the nuclear medical imaging procedure [4] and therefore either a more significant diagnosis or less exposure to the patient. Again, these improvements are strongly depending on the clinical task itself since the signal-to-background ratio is not independent from the region of investigation and the possible pathology.

12.4 Medical Physicists and Engineers Providing Tools for Optimized Spatial and or Temporal Resolution Compared to What Is Possible Due to Physics Limitations

The signal-to-background ratios will change over time as well as the concentration of the marker in the pathological tissues but also in tissues assumed to be healthy. The underlying pharmacokinetics might give more detailed insights into the pathology of the patient. However, the time resolution achievable with a suitable administered activity and correspondingly a suitable exposure to the patient is quite limited in nuclear medicine. This is also true for the spatial resolution of the nuclear medical imaging systems as it can be learnt from Fig. 12.1. For some patients or investigations, either one of those limitations can be problematic to deduce the underlying pathological reason for a clinical case. As an example it should be mentioned that for investigations in small children, it might be of big interest to improve the resolution of, e.g., PET imaging down to the inferior limit of about 1 mm, which is due to the path length of the positron before the annihilation process. This would allow a much better correlation to anatomical structures in children. In a similar way, high-resolution images might be beneficial in (minimal invasive) surgeries, e.g., in lymph node characterization during surgeries. A solution could be the addition of patient-near high-resolution detectors as designed by the MADEIRA consortium (as described in Chap. 11) or manually movable and positioned additional detectors. For SPECT, movable detectors could be placed on positions most relevant for reconstruction to improve, e.g., time resolution or handling during surgery procedures. This is a concept already on the market [5].

12.5 Mathematicians and Software Developers Contributing Algorithms and Software Packages Specifically Designed for the Development

New methods and intelligent algorithms are necessary to use the information gained from the radiation emitted from the patient either on the determined chords (PET) or in the single detector elements (SPECT) in the best suitable way. Meanwhile, this sometimes even includes a determination of the most wanted additional data to gain the highest ratio of additional information per scan time.

The most important problem of nuclear medical imaging is still to separate the noise from the wanted information content. This is in many aspects of great importance for the reconstruction process as well since without such a separation it is hard to find the subject of interest at all given the low amount of radiopharmaceutical typically needed to be seen and the administered activity being as low as possible. In addition, due to high noise in the raw data, there is always a great chance to introduce artifacts into the reconstructed images. This gets more difficult and prominent, if there are two different kinds of data sets, e.g., with various spatial resolution. A number of different efforts have been made to solve these problems in a best possible way. The most common and most important tool today is certainly the iterative reconstruction scheme [6–8]. One could also try to reduce noise before applying analytical reconstruction processes at least as a first step.

As it is valid for all other topics described before, it is necessary to optimize this process in connection with the other optimization steps. The reconstruction method to choose will depend on the physical detector and geometry configuration, the biodistribution in the patient but also on the clinical task (choosing between more noisy but higher resolved images and less noisy, less resolved once). The clinical task might also influence the last step in the imaging process, the image presentation for diagnosis. Here all kinds of linear and non-linear image processing methods can be used in principle. However, it has to be guaranteed that no useful information is hidden by trying to produce good-looking images.

One aspect that really belongs to all five optimization steps and which cannot be neglected in this list of optimization needs is the question of how to choose the right imaging technology. Assuming that the need for molecular imaging is clear and justified, the main question was whether the tracer that is available is a PET or a SPECT tracer. Since most systems already used today are hybrid systems offering anatomical and functional information (mostly SPECT/CT and PET/CT), the question will arise whether the combination of nuclear medical imaging with the anatomical imaging using CT is optimal or, e.g., combinations with MRI will be advantageous. As described in Chap. 4, there are already first systems available combining MR imaging with PET systems. From the point of view of the nuclear imaging component, it is first of all important to get information useful for attenuation correction. In a second step it is helpful for the diagnosis to combine the anatomical and the functional information. Depending on the clinical task, it has to be decided with which kind of combination these two aspects can be done best regarding the quality of the resulting diagnosis. It will most probably depend strongly on the body region and structures to be detected.

To end with a summary of this chapter, it can be said there will not be an optimal system for a given clinical task for the benefit of each patient if not all disciplines are working closely together.

References

1. ICRP Publication 103 – The 2007 recommendations of the International Commission on Radiological Protection. Ann ICRP 37(2–4)
2. Mattsson S, Hoeschen C (eds) (2012) Radiation protection in nuclear medicine. Springer, Berlin
3. http://www.world-nuclear.org/info/inf55.html. Accessed 26 Oct 2012
4. Janzen T, Tavola F, Giussani A et al (2010) Compartmental model of ^{18}F-choline. In: Molthen RC, Weaver JB (eds) Medical imaging 2010: biomedical applications in molecular, structural, and functional imaging. SPIE, Bellingham, WA
5. Valdés Olmos RA, Vidal-Sicart S, Nieweg OE (2010) Technological innovation in the sentinel node procedure: towards 3-D intraoperative imaging. Eur J Nucl Med Mol Imaging 37(8):1449–1451. doi:10.1007/s00259-010-1468-3
6. Küstner T, Weidendorfer J, Schirmer J, Klug T, Trinitis C, Ziegler SI (2009) Parallel MLEM on multicore architectures. In: Computational science – ICCS 2009, LNCS, vol 5544, pp 491–500
7. Rafecas M, Boning G, Pichler BJ, Lorenz E, Schwaiger M, Ziegler SI (2004) Effect of noise in the probability matrix used for statistical reconstruction of PET data. IEEE Trans Nucl Sci 51(1):149–156
8. Söderberg M, Mattsson S, Oddstig J, Uusijärvi-Lizana H, Valind S, Thorsson O, Garpered S, Prautzsch T, Tischenko O, Leide-Svegborn S (2012) Evaluation of image reconstruction methods for 123I-MIBG-SPECT: a rank-order study. Acta Radiol 53(7):778–784

Index

A. Giussani and C. Hoeschen (eds.), *Imaging in Nuclear Medicine*,
DOI 10.1007/978-3-642-31415-5, © Springer-Verlag Berlin Heidelberg 2013

MIX
Papier aus verantwortungsvollen Quellen
Paper from responsible sources
FSC® C105338

If you have any concerns about our products,
you can contact us on
ProductSafety@springernature.com

In case Publisher is established outside the EU,
the EU authorized representative is:
Springer Nature Customer Service Center GmbH
Europaplatz 3, 69115 Heidelberg, Germany

Printed by Libri Plureos GmbH
in Hamburg, Germany